Diet, Nutrition and Lifestyle in Aging and Age-Related Diseases

Diet, Nutrition and Lifestyle in Aging and Age-Related Diseases

Guest Editors

Emiliana Giacomello
Luana Toniolo

MDPI

Basel • Beijing • Wuhan • Barcelona • Belgrade • Novi Sad • Cluj • Manchester

Guest Editors

Emiliana Giacomello
Department of Medicine
University of Trieste
Trieste
Italy

Luana Toniolo
Department of Biomedical Sciences
University of Padova
Padova
Italy

Editorial Office
MDPI AG
Grosspeteranlage 5
4052 Basel, Switzerland

This is a reprint of the Special Issue, published open access by the journal *Nutrients* (ISSN 2072-6643), freely accessible at: https://www.mdpi.com/journal/nutrients/special_issues/Nutrition_Aging_Diseases.

For citation purposes, cite each article independently as indicated on the article page online and as indicated below:

Lastname, A.A.; Lastname, B.B. Article Title. *Journal Name* **Year**, *Volume Number*, Page Range.

ISBN 978-3-7258-6294-8 (Hbk)
ISBN 978-3-7258-6295-5 (PDF)
https://doi.org/10.3390/books978-3-7258-6295-5

Contents

About the Editors

Emiliana Giacomello

Emiliana Giacomello works as an Assistant Professor at the Department of Medicine, Surgery and Health at the University of Trieste. Her main interests are focused on the study of skeletal muscle plasticity under differentiation, development, and different physiological conditions, such as aging, bed rest, and obesity. In this context, her most recent research has been centered on studying the impact of diet and calorie restriction mimetics in animal models. Her research encompasses in vitro and vivo studies that include cell culture, animal models, and human tissues. She specializes in the characterization of skeletal muscle cells by histomorphology methodologies. She is an expert on the application of conventional and confocal fluorescence microscopy, which allows for collaborating with multiple groups in order to share and implement knowledge.

In parallel to her research activity, she fulfills didactic commitments in the courses of Medicine and Surgery and Technical Health Professions.

Luana Toniolo

Luana Toniolo works as Associate Professor in the muscle physiology and biophysics laboratory of the Department of Biomedical Sciences at the University of Padova as a member of the "Muscle Contractility and Plasticity" research group.

Her research activity is centered on muscle physiology, with emphasis on the cellular and molecular aspects of diversity among skeletal muscle fibers. She specializes in the preparation and manipulation of single-skinned cells and ex vivo muscle preparations for studying the physiological characteristics of mammalian skeletal muscle.

She is interested in various problems related to muscle tissue, particularly the physiological responses of skeletal muscle and its adaptations to stimuli such as training, immobilization, denervation, or its physiological changes such as aging.

Alongside her research activity, she is involved in various didactic commitments, mainly in the context of the degree course in Medicine and Surgery. Since 2011, she has been a member of the Italian Society of Physiology.

Preface

Aging entails the alteration of multiple mechanisms at the cell and organ level that ultimately contribute to the functional decline of an organism. As a result, older people experience a complex condition named frailty, which includes alterations of their physical and psychological abilities, comprising weakness, lower physical activity, worsening of psychological conditions, and low-grade chronic inflammation. Frailty has been demonstrated to be a tunable condition, and its onset and development is correlated with several factors such as socioeconomic background, nutritional habits, and physical exercise.

The current Issue was motivated by the increase in the older population and the need to deal with related problems. Recently, there has been significant improvement in sanitary, nutritional, and socioeconomic conditions, which has led to an increase in life expectancy. However, this lifespan extension has negative side effects consisting of age-related complications that compromise the quality of life of older individuals.

The present Issue, dedicated to professionals and translational researchers in the field of aging medicine, delves into the factors that regulate age-related diseases and frailty and reports on strategies that promote healthy aging, elucidating how factors connected to one's lifestyle, including physical exercise, environmental and economic conditions, and other factors, can modulate the quality of aging. The aim of this Issue is to contribute further knowledge to the mechanisms underlying frailty and age-related diseases and propose strategies to improve quality of life in older individuals.

Emiliana Giacomello and Luana Toniolo
Guest Editors

nutrients

MDPI

Article

Randomised Clinical Trial to Analyse the Efficacy of Eggshell Membrane to Improve Joint Functionality in Knee Osteoarthritis

Fernando Cánovas [1], María Salud Abellán-Ruíz [1], Ana María García-Muñoz [1], Antonio Jesús Luque-Rubia [1], Desirée Victoria-Montesinos [1], Silvia Pérez-Piñero [1], Maravilla Sánchez-Macarro [1] and Francisco Javier López-Román [1,2,*]

1 Health Science Department, Campus de los Jerónimos, Universidad Católica San Antonio de Murcia UCAM, 30107 Murcia, Spain; fcanovas@ucam.edu (F.C.); msabellan@ucam.edu (M.S.A.-R.); amgarcia13@ucam.edu (A.M.G.-M.); ajluque@ucam.edu (A.J.L.-R.); dvictoria@ucam.edu (D.V.-M.); sperez2@ucam.edu (S.P.-P.); msanchez4@ucam.edu (M.S.-M.)

2 Primary Care Research Group, Biomedical Research Institute of Murcia (IMIB-Arrixaca), 30120 Murcia, Spain

* Correspondence: jlroman@ucam.edu

Abstract: Osteoarthritis is a source of chronic pain and disability. Dietary supplements have been shown to be a more secure option than NSAIDS. Particularly, the eggshell membrane has demonstrated efficacy in relieving joint pain and stiffness. A clinical trial was designed in which three groups were assigned to two different doses of this supplement and compared to a placebo control group. The primary outcome variable was knee pain, which was assessed using a visual analogue scale. Secondary outcome variables were knee functional ability, quadriceps muscle strength (assessed by isometric and isokinetic trials), and quality of sleep. All groups showed a significant decrease in pain perception, although maximum values were obtained in the high-dose group. Isokinetic and isometric trials showed a significant increase in strength in the high-dose group. Eggshell membrane showed the potential to reduce pain and stiffness symptomatology. Here, for the first time, two quantitative variables (mobility and strength of knee joint) were used to accurately evaluate changes in the quality of life of subjects affected by knee joint pain. The results of this study indicate a dose-dependent response, which should be taken into account for later use in therapeutics to establish the correct dosage.

Keywords: knee pain; dietary supplement; stiffness; glycosaminoglycans

1. Introduction

Increasing life expectancy has led to structural changes in current human populations [1]. Ageing is posing a challenge for increasing care needs due to physical and mental multimorbidities in the 21st century [2] since evidence that longevity coincides with an extended period of good health is scarce [3]. In fact, ageing and increased body mass index (BMI) are leading causes of osteoarthritis (OA) [4].

OA is the most prevalent joint disorder and source of chronic pain and disability in developed countries [5,6]. This degenerative disease of the joint affects the locomotor system and is characterised by a loss of articular cartilage, as well as an osseocartilaginous proliferation of the subchondral and articular margins [7]. About 10% of adult people suffer some type of moderate to severe OA, and this percentage increases with age and is even more accentuated in those over 50 to 55 years old [8]. The hips and knees are the joints most affected by this disorder.

Acute pain is a commonly associated symptom; therefore, pharmacotherapy is indicated as treatment through the use of analgesics and non-steroidal, anti-inflammatory drugs (NSAIDS) [7,9]. However, most of these treatments have shown limited effectiveness and have induced several multiorgan toxicities [10]. Currently, dietary supplements are commonly used to counteract pain in a combination of hyaluronic acid, glucosamine, and chondroitin. However, some dietary supplements have also been examined [11,12] and have been shown to be a more secure option than NSAIDS [13].

Among them, eggshell membrane has demonstrated efficacy in relieving joint pain and stiffness [14–19]. The eggshell membrane is composed mainly of fibrous collagen proteins, types I, V, and X. It also contains glycosaminoglycans, such as chondroitin sulphate and dermatan sulphate, and hexosamines, such as glucosamine [20]. In addition, hyaluronic acid has been shown to be present in significant amounts [21]. Therefore, ESM has been evaluated as a possible treatment for OA as a natural source of an optimal combination of such compounds [17,19]. Despite the fact that some clinical trials have demonstrated promising results in improving the functionality of the knee joint [19], some lack of knowledge still remains regarding dosage.

Therefore, the objective of this study was to determine the efficacy of a food supplement extracted from the internal membrane of the eggshell on joint functionality (knee functional ability and quadriceps muscle strength) and perceived pain (visual analogic scale and quality of sleep) in individuals diagnosed with OA, after a consumption period of eight weeks. Two doses were tested in order to evaluate efficacy in a placebo-controlled clinical trial.

2. Materials and Methods

A randomised, controlled, double-blind, single-centre clinical trial was designed, in which three groups were assigned to different doses of an internal membrane, eggshell-based supplement (Figure 1). This study was conducted in 2018–2019 on 80 patients over 18 years of age with diagnosed knee osteoarthritis and chronic knee pain.

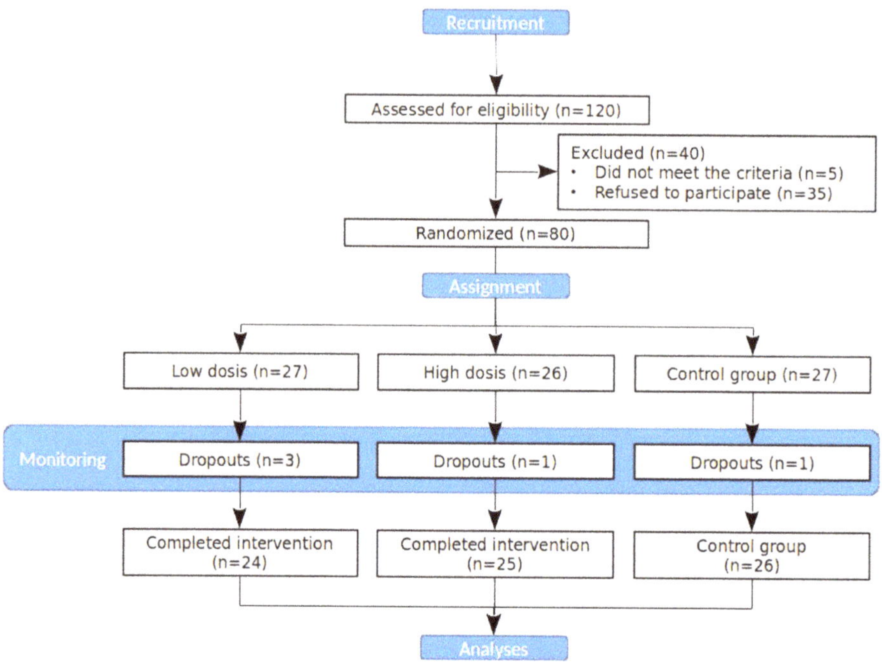

Figure 1. Flowchart representing the different stages of the trial. Randomisation profile shows flow of patients in the three groups: high-dose (500 mg), low-dose (300 mg), and control group.

The trial design followed CONSORT guidelines. Written, informed consent was signed by all participants after approval by the University's Institutional Science Ethics Committee. Inclusion criteria were as follows: subjects diagnosed with arthrosis with functional grades I–III according to criteria from the American College of Rheumatology, Atlanta; persistent knee pain associated with this pathology, with initial punctuation of at least 30 mm in a visual analogue scale of 10 cm; and absence of chronic treatment with narcotics, NSAIDs, or immunosuppressants. Exclusion criteria were as follows: terminal disease; known chronic inflammatory diseases affecting the musculoskeletal system (rheumatoid arthritis, gout, pseudo-gout, Paget's disease, chronic pain syndrome, etc.); other serious illnesses that would limit the execution of aerobic or resistance exercises (musculoskeletal conditions, limiting pneumopathy, presence of arrhythmia); body mass index (BMI) above 32; subjects who, at the time of the study, were being treated with glucosamine, chondroitin sulphate, collagen, or hyaluronic acid infiltrations, or consuming any supplement indicated for joint health; subjects under pharmacotherapy (narcotic drugs, steroid anti-inflammatory drugs, or immunosuppressants); known allergy to eggs; pregnant or breastfeeding women; and incapacity to understand informed consent.

Treatment consisted of once-daily oral ingestion of 300 mg (low dose) and 500 mg (high dose) of ESM® eggshell membrane (Torolis Explotaciones, S.L., Navarra, Spain) in vegetarian capsules that were stored in closed containers at ambient temperature. The control group consumed a placebo based on encapsulated maltodextrin matching the weight and aspect of treatment. Eggshell membrane extract was composed mainly of proteins (>90%), fibrous collagen types I, V, and X (<13%), and elastin (<4–5%). The extract also contained glycosaminoglycans, such as chondroitin sulphate (<2%); dermatan and keratin sulphates (<1%); and hexosamines, such as glucosamine (<2%), hyaluronic acid (<2%), IGF-1 (12 ng/g), TGF-beta (0.75–7.23 ng/g), calcitonin (10–25 ng/g), and progesterone (0.3–0.33 ng/g). Clinic visits were scheduled for subjects at study initiation and at the end of 8 weeks following the onset of treatment. Compliance with the treatment was evaluated during the last visit via patient interview and by counting the number of unused capsules still remaining. Lifestyle variables included a dietary recall interview of the previous 3 days before starting the nutritional supplement and the last 72 h before the end of the 8-week consumption of the product. The level of physical activity was recorded using the Global Physical Activity Questionnaire (GPAQ), and results were expressed as MET-min/week. Specific rules were given to subjects about maintaining the same level of physical exercise during the trial. Weight was registered via bioimpedance.

The primary outcome variable was knee pain which was assessed using a visual analogue scale (VAS) of 10 cm, defining extreme limits such as 'no pain at all' and 'pain as bad as it could be' [22]. Values lower than 4 in VAS meant mild to moderate pain, a value between 4 and 6 implied the presence of moderate to severe pain, and pain with values higher than 6 implied the presence of very intense pain [22]. This scale was recorded before and after strength assessment at both baseline and final visits. VAS scale was also recorded throughout the 8 weeks at the time the participant awoke each morning through the completion of a diary. Adverse events were also recorded.

Secondary outcome variables were knee functional ability, quadriceps muscle strength (assessed by isometric and isokinetic trials), and quality of sleep.

Functional ability was evaluated using the Western Ontario and McMaster Universities Osteoarthritis (WOMAC) index, through an adapted Spanish version of the WOMAC questionnaire [23]. This questionnaire determined pain and functional ability. Within the questionnaire, there are three subscales, 5 questions referring to joint pain, 2 questions on stiffness, and 17 on functional capacity. Each question is graded from 0 to 4, with 0 meaning 'none' and 4 meaning 'very much'. The WOMAC questionnaire had to be completed before each of the strength assessments at both visits.

Muscle strength was determined using a Biodex System 3 Dynamometer (Biodex Medical System, Shirley, New York, NY, USA) in order to assess isometric (straining your muscles without moving or bending your joints) and isokinetic (performed at a consistent speed, which can be increased as you progress) strength of the knee flexors and extensors. It is important to note that changes in strength should be associated with changes in perceived pain and not with changes in the physical condition of the muscle. Trials involved two maximum isokinetic and two continuous maximum isometric repetitions at 60°/s and 90°/s on their right leg, respectively. Isokinetic trials collected information about muscle peak torque (PT, i.e., the maximum force that a muscle group can produce); total work of the maximum repetition (TWMR, i.e., maximum strength exerted at the time of making the movement); and total work (TW, i.e., strength exerted during all repetitions to cause the displacement of the leg and overcome the resistance offered by the isokinetic dynamometer protocol). Isometric trials assessed muscle peak torque (PT/BW) and maximum average peak torque (MAPT, i.e., medium torque). Verbal encouragement was provided during the tests, and adequate rest and recovery times were provided between contractions in order to minimise fatigue.

Quality of sleep was assessed by the Pittsburgh Sleep Quality Index (PSQI) [24], which has shown good psychometric properties and validity for application in the adult population. PSQI evaluated seven domains: subjective sleep quality, sleep latency, sleep duration, habitual sleep efficiency, sleep disturbances, use of sleeping medication, and daytime dysfunction over the past month. Each subject self-rated each of these seven areas of sleep. Each domain is scored based on a 0–3 on the Likert scale: 0 (very good), 1 (good), 2 (poor), and 3 (very bad). The PSQI score was calculated as the sum of the scores, which varied between 0 and 21. This test was filled in before each of the strength evaluations at each visit.

ANOVA with repeated measures was used to compare variables that were obtained at the beginning and end of this trial. All variables were checked for normality. Homogeneity of variables among groups was also checked in the baseline in order to avoid confounding variables. Bonferroni correction was used for all comparisons between intervention groups and control. Type I error rate was set at $\alpha = 0.05$. Analyses were performed by using SPSS statistical software v.21.0 (IBM Corp., Armonk, NY, USA, EEUU).

3. Results

A total of 80 out of 120 subjects assessed for eligibility were randomised (see Figure 1 for a trial flow), from which 75 (93.75%) completed this study (36 men and 39 women, mean age 38.40 ± 13.54 years old). Five individuals did not match the criteria for inclusion in the trial, and thirty-five individuals refused to participate. Demography information is provided for each group in Table 1. The placebo group included a total of 26 participants with a mean age of 41 ± 14.36 years old (10 men and 16 women). The low-dose group showed an average age of 37.38% ± 12.29 years old (12 men and 12 women). The high-dose group had an average age of 36.36 ± 13.54 years old (14 men and 11 women). At the beginning of this study, the percentage of body fat in the control group was 27.75 ± 1.7%, while for experimental groups, it was 26.89 ± 1.7% (300 mg) and 24.39 ± 1.7% (500 mg). By the end, the control group registered a body fat percentage of 28.32 ± 1.7%, while the 300 mg and 500 mg consumption groups registered a body fat percentage of 25.98 ± 1.7% and 24.82 ± 1.7%, respectively. No significant differences were found. BMI (normal vs. overweight/obesity) did not show significant differences between groups, as well as between initial and final conditions after ending of this study.

At the beginning of this study, subjects indicated self-medication with NSAIDs in 18.5%, 14.8%, and 19.2% of control, low-, and high-dose experimental groups, respectively. Although subjects did not report changes in such medication, diary annotations recorded punctual pain medication in 25% of them.

Table 1. Demography characteristics of the 75 subjects who finished this study by group.

	Control (n = 26)	Exp 300 (n = 24)	Exp 500 (n = 25)	Total
Age (mean ± SD)	41.31 ± 14.36	37.38 ± 12.29	36.36 ± 13.54	38.44 ± 13.54
BMI (mean ± SD)	25.3 ± 4.0	25.4 ± 4.0	24.6 ± 3.0	25.1 ± 3.6
BMI < 25; n (%)	12 (46.2%)	10 (41.7%)	15 (60%)	37 (49.3%)
BMI ≥ 25; n (%)	14 (53.8%)	14 (58.3%)	10 (40%)	38 (50.7%)
Sex; n (%)				
Women	16 (61.5%)	12 (50.0%)	11 (44.0%)	39 (52.0%)
Men	10 (38.5%)	12 (50%)	14 (56.0%)	36 (48.0%)

All three groups showed a statistically significant decrease in pain perception by the end of this study (Table 2). Furthermore, there were significant differences between groups, as the high-dose group showed a statistically significant reduction in pain with respect to the control group (Table 2).

Table 2. Visual analogue scale (VAS) for perceived pain, functional capacity by means of the WOMAC test, and Pittsburgh Sleep Quality Index (PSQI) for each of the visits and doses. Δ represents the increment from the start to the end of this trial. Significance level for the differences between values at the beginning and end of this trial.

		Control	Low Dose	High Dose
VAS (p = 0.001)	**Initial**	4.90 (1.48)	4.67 (1.37)	5.42 (1.48)
	Final	3.64 (1.40) ***	2.25 (1.66) ***	1.90 (1.78) ***
	Δ	−1.269	−2.417	−3.52
WOMAC (p = 0.183)	**Initial**	25.96 (13.42)	26.29 (15.81)	24.80 (10.92)
	Final	20.32 (13.17) **	16.04 (12.28) ***	14.52 (10.20) ***
	Δ	−5.64	−10.25	−10.28
PSQI (p = 0.301)	**Initial**	6.77 (2.82)	6.00 (2.34)	6.60 (3.76)
	Final	6.27 (3.14)	5.54 (2.89)	5.00 (3.03) **
	Δ	−0.5	−0.46	−1.6

Significance levels: *** < 0.001 < ** < 0.01 < * < 0.05.

The evolution of pain perception as soon as a subject awoke during the study period can be also followed in Table 3. No differences were observed between groups in the baseline ($p > 0.1$), registering 4.75 ± 1.16 points (control), 4.69 ± 1.18 points (low dose), and 5.23 ± 1.32 points (high dose). A progressive decrease in the evolution of pain can be observed for all groups, and the lowest values on the scale were achieved in the last week for all groups (Table 3). The control group showed mild to moderate perceived pain by the end of this study (3.91 ± 2.16 points), although this decrease (1.27 points) was not significant ($p > 0.05$). The other two groups registered a significant decrease in perceived pain, decreasing to 3.13 ± 1.43 points (mild to moderate pain) in the low-dose group, and decreasing to 2.98 ± 1.51 points (mild pain) in the high-dose group (Table 3). Comparison of the evolution among the three study groups showed significant differences, and the high-dose group significantly reduced perceived pain with respect to the control group ($p < 0.014$).

The WOMAC results were homogeneous among the three groups at the beginning of this trial (Table 2). The control group started this study with values of 25.96 ± 2.7 points, while the low- and high-dose groups started with similar values of 26.29 ± 2.7 and 24.80 ± 2.7 points, respectively ($p > 0.1$). Despite the fact that all three groups showed a significant decrease in the WOMAC scale score (Table 2), both groups that consumed this product showed a more pronounced decrease, and therefore an improvement in functional capacity and quality of life (Table 2).

Table 3. Weekly mean (standard deviation) of visual analogue scale (VAS) for perceived pain during the period of study. Δ represents the increment from the start to the end of this trial. Significance level for the differences between values at the beginning and end of this trial.

	Week1	Week2	Week3	Week4	Week5	Week6	Week7	Week8	Week9	Δ
Control	4.75 (1.16)	4.74 (1.72)	4.62 (1.93)	4.52 (2.27)	4.43 (1.88)	4.31 (2.27)	4.13 (2.21)	4.19 (2.29)	3.91 (2.16)	−0.835
Low dose	4.69 (1.18)	4.37 (1.45)	3.99 (1.47)	3.77 (1.41)	3.89 (1.17)	3.53 (1.41)	3.62 (1.42)	3.32 (1.57)	3.13 (1.43)	−1.559
High dose	5.23 (1.32)	4.93 (1.55)	4.3 (1.74)	4.08 (1.7)	3.84 (1.68)	3.65 (1.81)	3.4 (1.71)	3.23 (1.59)	2.98 (1.51)	−2.250 ***

Significance levels: *** < 0.001 < ** < 0.01 < * < 0.05.

Isokinetic and isometric trials showed no significant differences at the beginning of this study for all variables included (Table 4). Although an increase in strength, measured as PT, TWMR, and TW, with respect to baseline values, was observed in those groups subjected to consumption of eggshell membrane product at both doses, the decrease in pain was only significant in the high-dose group for all the variables studied here, accounting for 7.2, 10.04 and 47.42 N × m, respectively (Table 4). The control group registered a decrease in strength for two variables (PT = −1.48 N x m and TW = −3.21 N × m), although such results lacked statistical significance (Table 3). Similar results were found for isometric trials. The high-dose group showed a significant increase in strength for both PT (25.18 N × m) and MAPT (20.76 N × m). The low-dose group showed an increase in PT (20.36 N × m) and MAPT (16.29 N × m), although the latter lacked statistical significance (Table 4). PT registered non-significant results for the control group that showed a slight decrease (−0.41 N × m), and MAPT registered a slight increase that also lacked significance (1.01 N × m).

Table 4. Isokinetic trials variables at 60°/s: muscle peak torque (PT), total work of the maximum repetition (TWMR), and total work (TW). Isometric trials assessed at 90°/s: muscle peak torque (PT) and maximum average peak torque (MAPT). Δ represents the increment from the start to the end of this trial. Significance level for the differences between values at the beginning and end of this trial. p values indicate significance level for comparison of groups by the duration of this trial for each variable. Units were measured in N × m.

		Isokinetic at 60°/s		
		Control	**Low Dose**	**High Dose**
PT (p = 0.048)	**Initial**	57.93 (22.36)	62.15 (24.28)	59.28 (26.52)
	Final	56.45 (22.66)	62.98 (24.92)	66.48 (28.23)
	Δ	−1.48	0.83	7.2 ***
TWMR (p = 0.016)	**Initial**	64.60 (28.48)	68.91 (30.40)	66.89 (31.94)
	Final	64.66 (27.73)	70.53 (30.17)	76.93 (35.08)
	Δ	0.06	1.62	10.04 ***
TW (p = 0.017)	**Initial**	294.89 (137.08)	309.38 (150.72)	297.72 (147.07)
	Final	291.68 (130.01)	322.61 (144.70)	345.13 (167.64)
	Δ	−3.21	13.23	47.42 ***
		Isometric at 90°/s		
		Control	**Low Dose**	**High Dose**
PT (p = 0.018)	**Initial**	145.41 (61.07)	137.99 (61.16)	146.73 (70.07)
	Final	145.00 (61.42)	158.35 (56.72)	171.92 (84.64)
	Δ	−0.41	20.36 **	25.18 ***
MAPT (p = 0.016)	**Initial**	138.16 (58.35)	137.40 (52.22)	141.27 (57.34)
	Final	139.17 (59.23)	153.69 (52.52)	161.04 (78.39)
	Δ	1.01	16.29	20.76 ***

Significance levels: *** < 0.001 < ** < 0.01 < * < 0.05.

Pittsburgh test scores provided similar scores at the beginning of this trial for all three groups (Table 2), and no significant differences were found between them. At the end of the study, the control group kept the same scores. Both treated groups showed a decrease, though this was higher in the high-dose group (1.6 points) than in the low-dose group

(0.46 points). These differences observed in the high-dose group were significantly different from those obtained for the other two groups, indicating an improvement in the quality of sleep for the high-dose group compared with the other two groups.

Subjects did not report any adverse events related to the consumption of the product administered.

4. Discussion

Pain due to osteoarthritis is known to induce long-term consequences on health as a consequence of a lack of exercise [16]. In fact, other comorbidities, such as being overweight or ageing, could aggravate these consequences and could influence other ones as a result of immobility due to pain.

Eggshell membrane showed the potential to reduce such symptomatology. Although all of the groups registered a decrease in perceived pain, both doses used in this study induced a bigger reduction than in the control group, indicating a relationship between eggshell membrane consumption and reduction in pain. Such results are compatible with previous studies that demonstrated the efficacy of this nutraceutical to alleviate knee pain, symptomatology, and joint functionality [14–19]. These results are higher than those registered for other supplements such as curcuma [12]. However, dose-dependent efficacy was registered in our study (Table 2). Despite the fact that all groups showed a reduction in pain, objective measures of improvement in functional capacity and sleep quality support such a hypothesis. A 500 mg dose induced a reduction in perceived pain of 1.1 points more than a 300 mg dose. Furthermore, a slight reduction in functional capacity and an improvement in sleep quality were also registered for high vs. low doses. All three signs together indicate an improvement in knee pain, although an increase in the dose used does not show a proportional effect in these variables.

Reduction in perceived pain was decremental and was observed during the entire study period, as registered by the information collected as soon as the subject awoke in the morning. Such a variable registered a significant weekly evolution of pain through the study for the high dose, showing a final reduction of 0.69 points more than the low dose (Table 3). Those results are also in agreement with the trials performed in this study. The WOMAC scale score and quality of sleep index (Table 2) showed a decrease in both groups of consumers, therefore registering an improvement in functional capacity and quality of life, although differences between doses were not significant. This is probably due to the fact that a decrease in pain and a recovery of knee joint functionality perception could make those two variables (functional capacity and quality of life) improve but only up to a certain level due to the subjectivity of tests. Perception of improvement could probably be achieved in both dosages, only showing slight differences between dosages registered by the tests used.

In fact, such an observation could be in agreement with the results obtained for muscle strength. All variables studied here registered an increase in both consumption groups, although a higher dose-dependent effect was registered than for other tests (Table 4). Isokinetic trial variables at 60°/s (PT, TWMR, and TW) show the highest differences: between four and ten times greater for the higher dose. Isometric trials assessed at 90°/s (PT and MAPT) also show significant differences between control and treatment groups, but with fewer differences between the low and high doses. These results show a significant improvement in muscle strength as a result of a reduction in pain perception. Moreover, such an effect is dose-dependent. Such a conclusion can make a difference when applying the use of eggshell membranes in therapeutics regarding knee pain; therefore, perceived pain reduction and functionality increase do not entirely reflect all of the benefits of consuming eggshell membranes. This could probably be due to the objectivity of muscle strength assessment through trials compared with pain and functionality tests, which would indicate a physiological improvement and, therefore, a more objective observation about the outcome of the use of eggshell membranes to treat knee pain.

Although previous studies addressed the study of the efficacy of eggshell membrane as a supplement to alleviate knee joint pain [14–18], here, for the first time, two quantitative variables (mobility and strength of knee joint), as well as qualitative ones (pain, functional ability, and sleep quality), were monitored to accurately evaluate changes in the quality of life of subjects affected by this common pathology. Such a variable set extends the findings of previous studies that have already observed changes in the functionality of knee joints by measuring the range of moti- on [19]. This study reports an improvement in all of the variables monitored in those subjects who were consuming eggshell membrane, most likely resulting from physiological changes in the musculature associated with the amelioration of knee functionality and a better quality of life due to a reduction in perceived pain. Moreover, our results also indicate a dose-dependent response, which should be taken into account for later use in therapeutics to establish the correct dosage.

Author Contributions: Conceptualisation, F.J.L.-R.; methodology, M.S.A.-R., A.J.L.-R., D.V.-M., S.P.-P. and A.M.G.-M.; validation, M.S.A.-R. and A.J.L.-R.; formal analysis, M.S.-M. and F.J.L.-R.; investigation, S.P.-P., D.V.-M. and M.S.A.-R.; resources, S.P.-P.; data curation, A.M.G.-M.; writing—original draft preparation, F.C.; writing—review and editing, F.C. and F.J.L.-R.; supervision, F.J.L.-R.; project administration, F.C.; funding acquisition, F.J.L.-R. All authors have read and agreed to the published version of the manuscript.

Funding: This study was performed at UCAM Catholic University of Murcia (Spain) and was sponsored by Torolis Explotaciones S.L, (Spain).

Institutional Review Board Statement: This study was conducted in accordance with the Declaration of Helsinki and approved by the Ethics Committee) of the Catholic University of Murcia (protocol code CE031808).

Informed Consent Statement: Informed consent was obtained from all subjects involved in the study.

Data Availability Statement: Not applicable.

Acknowledgments: The study product (ESM®) was provided by Torolis Explotaciones S.L. (Spain).

Conflicts of Interest: The authors declare no conflict of interest. The funders had no role in the design of the study; in the collection, analyses, or interpretation of data; in the writing of the manuscript; or in the decision to publish the results.

References

1. Griffin, J.P. Changing life expectancy throughout history. *J. R. Soc. Med.* **2008**, *101*, 577. [CrossRef] [PubMed]
2. The Lancet Public Health. Ageing: A 21st century public health challenge? *Lancet Publ. Health* **2017**, *2*, e297. [CrossRef]
3. Beard, J.R.; Officer, A.; de Carvalho, I.A.; Sadana, R.; Pot, A.M.; Michel, J.P.; Lloyd-Sherlock, P.; Epping-Jordan, J.E.; Peeters, G.; Mahanani, W.R.; et al. The World report on ageing and health: A policy framework for healthy ageing. *Lancet* **2016**, *387*, 2145–2154. [CrossRef]
4. Wallace, I.J.; Worthington, S.; Felson, D.T.; Jurmain, R.D.; Wren, K.T.; Maijanen, H.; Woods, R.J.; Lieberman, D.E. Knee osteoarthritis has doubled in prevalence since the mid-20th century. *Proc. Natl. Acad. Sci. USA* **2017**, *114*, 9332–9336. [CrossRef]
5. Cross, M.; Smith, E.; Hoy, D.; Nolte, S.; Ackerman, I.; Fransen, M.; Bridgett, L.; Williams, S.; Guillemin, F.; Hill, C.L.; et al. The global burden of hip and knee osteoarthritis: Estimates from the global burden of disease 2010 study. *Ann. Rheum. Dis.* **2014**, *73*, 1323–1330. [CrossRef]
6. Rejas-Gutierrez, J.; Llopart-Carles, N.; García-López, S.; Darbà, J. Disease burden on healthcare by pain severity and usual analgesic treatment in patients with symptomatic osteoarthritis: A Spanish nationwide health survey. *Expert Rev. Pharmacoecon. Outcomes Res.* **2021**, *21*, 711–719. [CrossRef]
7. Hussain, S.M.; Neilly, D.W.; Baliga, S.; Patil, S.; Meek, R. Knee osteoarthritis: A review of management options. *Scott. Med. J.* **2016**, *61*, 7–16. [CrossRef]
8. Bannuru, R.R.; Osani, M.C.; Vaysbrot, E.E.; Arden, N.K.; Bennell, K.; Bierma-Zeinstra, S.; Kraus, V.B.; Lohmander, L.S.; Abbott, J.H.; Bhandari, M.; et al. OARSI guidelines for the non-surgical management of knee, hip, and polyarticular osteoarthritis. *Osteoarthr. Cartil.* **2019**, *27*, 1578–1589. [CrossRef]
9. Hermann, W.; Lambova, S.; Muller-Ladner, U. Current treatment options for osteoarthritis. *Curr. Rheumatol. Rev.* **2018**, *14*, 108–116. [CrossRef]
10. Bindu, S.; Mazumder, S.; Bandyopadhyay, U. Non-steroidal anti-inflammatory drugs (NSAIDs) and organ damage: A current perspective. *Biochem. Pharmacol.* **2020**, *180*, 114147. [CrossRef]

11. Li, X.Z.; Zhang, S.N. Recent advance in treatment of osteoarthritis by bioactive components from herbal medicine. *Chin. Med.* **2020**, *15*, 80. [CrossRef] [PubMed]
12. Perkins, K.; Sahy, W.; Beckett, R.D. Efficacy of curcuma for treatment of osteoarthritis. *Evid. Based Complement Alternat. Med.* **2017**, *22*, 156–165. [CrossRef] [PubMed]
13. Shep, D.; Khanwelkar, C.; Gade, P.; Karad, S. Safety and efficacy of curcumin versus diclofenac in knee osteoarthritis: A randomized open-label parallel-arm study. *Trials* **2019**, *20*, 214. [CrossRef] [PubMed]
14. Brunello, E.; Masini, A. NEM brand eggshell membrane effective in the treatment of pain and stiffness associated with osteoarthritis of the knee in an italian study population. *Int. J. Clin. Med.* **2016**, *7*, 169–175. [CrossRef]
15. Ruff, K.J.; DeVore, D.P.; Leu, M.D.; Robinson, M.A. Eggshell membrane: A possible new natural therapeutic for joint and connective tissue disorders. Results from two open-label human clinical studies. *Clin. Interv. Aging* **2019**, *4*, 235–240.
16. Ruff, K.J.; Winkler, A.; Jackson, R.W.; DeVore, D.P.; Ritz, B.W. Eggshell membrane in the treatment of pain and stiffness from osteoarthritis of the knee: A randomized, multicenter, double-blind, placebo-controlled clinical study. *Clin. Rheum.* **2009**, *28*, 907–914. [CrossRef]
17. Danesch, U.; Seybold, M.; Rittinghausen, R.; Treibel, W.; Bitterlich, N. NEM® brand eggshell membrane effective in the treatment of pain associated with knee and hip osteoarthritis: Results from a six-center, open-label german clinical study. *J. Arthritis* **2014**, *3*, 136. [CrossRef]
18. Hewlings, S.; Kalman, D.; Schneider, L.V. A randomized, double-blind, placebo-controlled, prospective clinical trial evaluating water-soluble chicken eggshell membrane for improvement in joint health in adults with knee osteoarthritis. *J. Med. Food* **2019**, *22*, 875–884. [CrossRef]
19. Kiers, J.L.; Bult, J. Mildly processed natural eggshell membrane alleviates joint pain associated with osteoarthritis of the knee: A randomized double-blind placebo-controlled study. *J. Med. Food* **2021**, *24*, 292–298. [CrossRef]
20. Nakano, T.; Ikawa, N.I.; Ozimek, L. Chemical composition of chicken eggshell and shell membranes. *Poult. Sci.* **2003**, *82*, 510–514. [CrossRef]
21. Ha, Y.W.; Son, M.J.; Yun, K.S.; Kim, Y.S. Relationship between eggshell strength and keratan sulfate of eggshell membranes. Comparative biochemistry and physiology. *Part A Mol. Integr. Phys.* **2007**, *147*, 1109–1115. [CrossRef] [PubMed]
22. Haefeli, M.; Elfering, A. Pain assessment. *Eur. Spine J.* **2006**, *15* (Suppl. 1), S17–S24. [CrossRef] [PubMed]
23. Villanueva, I.; Guzman, I.M.M.; Toyos, F.J.; Ariza-Ariza, R.; Navarro, F. Relative efficiency and validity properties of a visual analogue vs a categorical scaled version of the Western Ontario and McMaster Universities Osteoarthritis (WOMAC) Index: Spanish versions. *Osteoarthr. Cartil.* **2004**, *12*, 225–231. [CrossRef] [PubMed]
24. Buysse, D.J.; Reynolds, C.F., 3rd; Monk, T.H.; Berman, S.R.; Kupfer, D.J. The Pittsburgh sleep quality index: A new instrument for psychiatric practice and research. *Psychiatry Res.* **1989**, *28*, 193–213. [CrossRef]

nutrients

MDPI

Article

The Combined Effects of Dietary Diversity and Frailty on Mortality in Older Taiwanese People

Wei-Ching Huang [1,2], Yi-Chen Huang [3], Meei-Shyuan Lee [1,4], Jia-Yau Doong [5], Wen-Harn Pan [6] and Hsing-Yi Chang [1,2,*]

[1] Graduate Institute of Life Sciences, National Defense Medical Center, Taipei 114201, Taiwan
[2] Institute of Population Health Sciences, National Health Research Institutes, Miaoli 35053, Taiwan
[3] Department of Nutrition, China Medical University, Taichung 406040, Taiwan
[4] School of Public Health, National Defense Medical Center, Taipei 114201, Taiwan
[5] Department of Nutritional Science, Fu Jen Catholic University, New Taipei City 242062, Taiwan
[6] Institute of Biomedical Sciences, Academia Sinica, Taipei 11529, Taiwan
* Correspondence: hsingyi@nhri.org.tw; Tel.: +886-037-206-166 (ext. 36333)

Abstract: Objective: To assess the prospective association between frailty and dietary diversity on mortality. Method: This prospective cohort study used the 2005–2008 Nutrition and Health Survey in Taiwan (N = 330; age ≥ 65 years) and this was linked to the Death Registry where we used the data that was recorded up to 31 January 2020. Dietary intake information was assessed using a 24-h dietary recall and food-frequency questionnaire, which were calculated a dietary diversity score (DDS; range, 0–6) and food consumption frequency. Assessment of frailty phenotypes was based on FRAIL scale which was proposed by the International Academy on Nutrition and Aging. Results: Frail older adults had a higher risk of all-cause mortality when they were compared to those with robust physiologies (hazard ratio [HR]: 3.73, 95% confidence interval [CI]: 2.13–6.52). Frailty and a lower DDS were associated with a higher risk of mortality (joint adjusted HR: 2.30, 95% CI: 1.11–4.75) which, compared with a robust physiology and higher DDS, were associated with a lower risk of mortality. Conclusions: Frailty and a lower DDS were associated with a higher mortality. Prefrailty and frailty with a higher DDS were associated with a lower risk of mortality when compared with those with prefrailty and frailty and a lower DDS. These results suggest that eating a wide variety of foods might reduce the risk of mortality in older adults with prefrailty and frailty.

Keywords: frailty; dietary diversity; mortality; older adults; Nutrition and Health Survey in Taiwan (NAHSIT)

1. Introduction

Frailty is a medical and complicated geriatric syndrome that is characterized by multisystem decline that is associated with decreased functional reserves and increased vulnerability with aging [1]. Most studies define physical frailty by unintentional weight loss, exhaustion, low physical activity, slowness, and low grip strength [2]. The possible causes of frailty include physiological, genetic, environmental, and nutritional factors [1]. The results from a systematic review and meta-analysis reported that the prevalence of frailty and prefrailty increases with age and it appears higher prevalence in upper middle-income countries, among community-dwelling older adults [3]. With the rapid growth of the older population, worldwide, frailty requires attention because those with this condition are at a higher risk of falling, developing a disability, hospitalization, and mortality, and they require more medical services [4].

Eating a healthy diet is an important non-pharmacological strategy for not only preventing and improving the variety of non-communicable diseases but also promoting independence, quality of life, and ultimately healthy aging [5,6]. Studies have explored dietary patterns [7,8] and dietary quality and how they might affect health or mortality [9].

A brief dietary measurement method that can predict general health or mortality would be useful for public health nutrition programs and in a community-dwelling setting. The dietary diversity score (DDS) is a dietary quality measurement method, which is a simple and rapid tool that requires no devices or complicated measurements [9,10]. Older adults with more dietary diversity usually have a better dietary nutrient and energy intake profile as well as a lower mortality [9].

Our previous study showed that older adults with frailty had a lower dietary diversity, and those with both had a higher risk of developing worsening cognitive function [11]. A Taiwanese cross-sectional study demonstrated that a dietary pattern with more fruit, nuts and seeds, tea, vegetables, whole grains, shellfish, milk/yogurt/cheese, and fish was associated with a reduced prevalence of frailty [12]. It is possible that the cause of frailty might not just be due to the inadequate intake of individual nutrients or foods. Older adults with a higher dietary diversity may have a reduced mortality that is related to cognitive impairment [13], poor appetite [14], and diabetes mellitus [15]. Moreover, the results from a prospective cohort study demonstrated that a poor diet quality may increase the incidence of frailty [16], subsequently leading to a higher risk of mortality [17].

It is unknown whether dietary diversity can reduce the mortality that is associated with frailty. Therefore, the aim of this study was to investigate the prospective associations of dietary diversity and frailty severity to all-cause mortality in older adults.

2. Materials and Methods

2.1. Participants' Data Sources

The data were obtained from the Nutrition and Health Survey in Taiwan (NAHSIT) 2005–2008; detailed information on the design and methods of this survey has been published elsewhere [18]. A total of 968 older adults (aged $\geq$ 65 years) completed the household interviews. There were 330 participants with a national identification (ID) number that could be used to determine their survival status (Figure 1); participants were followed up for a median of 11.7 years until 31 January 2020. All participants provided written informed consent, and this study was approved by the Research Ethics Committee of the National Health Research Institute (EC1080701-E; Miaoli, Taiwan).

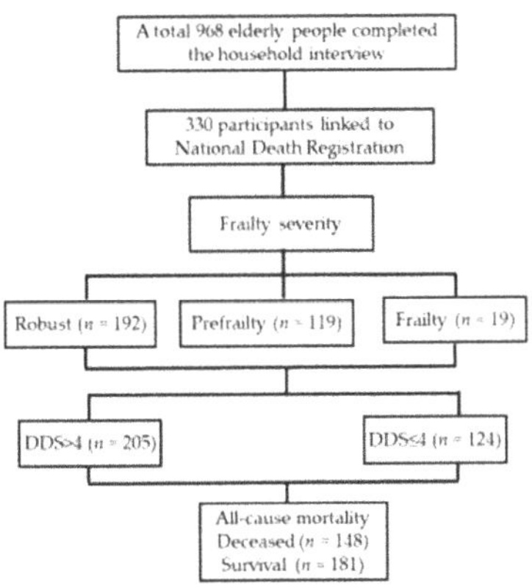

Figure 1. Flow chart for participant selection.

2.2. Frailty

Frailty and prefrailty were assessed using the FRAIL scale to assess frailty which was proposed by the International Academy on Nutrition and Aging [19]. It has been validated in previous studies [20]. The FRAIL scale includes five criteria: fatigue, resistance, ambulation, illnesses, and loss of weight [19]. Fatigue was measured by asking: "how much of the time during the past four weeks have participants felt tired?" The responses were measured from 1 (never) to 6 (all the time). Answers of 5 (most of the time) or 6 (all the time) were scored as 1; the other answers were scored as 0. Resistance was assessed by asking the participants how difficult it was to climb a flight of stairs. Responses of "Yes, limited a little or a lot" were scored as 1. Ambulation was assessed by asking the participants about the difficulty that they have walking a one block distance. Responses of "Yes, limited a lot" or "Yes, limited a little" were given a score of 1; the other responses were scored as 0. Illnesses were assessed by asking the participants if a medical professional had ever diagnosed them with any of the following illnesses: hypertension, cancer, diabetes, chronic lung disease, heart disease, asthma, arthritis, stroke, or kidney disease. A report of five or more illnesses received a score of 1. Weight loss was assessed by asking the participants if they had lost >5% of their weight in the past year. However, there was no equivalent variable that was available from the 2005–2008 NAHSIT database. Therefore, in this study, loss of weight was assessed by the participants who reported unintentional weight loss in the past month. Responses of "Yes" were given a score of 1. FRAIL scores ranged from 0 to 5, and these indicated frail (3–5), prefrail (1–2), and robust (0) conditions.

2.3. Dietary Intake Information

Dietary information was obtained by issuing the 24-h dietary recall and simplified food frequency questionnaire (FFQ). The 24-h dietary recall was used to estimate the nutrient intake, and the data were used to calculate the DDS [21]. The FFQ contains questions on the frequency of the consumption of 60 main food items per month/week/day during the month prior to taking the questionnaire. The validation of a similar simplified FFQ was previously reported. The Spearman rank correlation coefficients between the 24-h dietary recall and FFQ data ranged from 0.132 (whole grain) to 0.678 (dairy) for men, and −0.052 (whole grain) to 0.759 (dairy) for women [22].

Dietary quality was assessed using the DDS [10]. It was derived from the 24-h dietary recall of six food groups including dairy, soybeans/fish/eggs/meat, grain, fruit, vegetables, and oil/fat/nuts, in accordance with the Taiwanese Food Guides, and yielded a total score of 6 points. Achieving half of the recommended serving numbers per day of the six groups was required for a score of one [9]. The method that was used to estimate the serving numbers of the six food groups is provided in the Table S1. A DDS of >4 and $\leq$4 were considered high and low dietary diversity [9], respectively.

2.4. Mortality

The National Death Registry data were obtained from the Health and Welfare Data Science Center, Ministry of Health and Welfare. The 2005–2008 NAHSIT dataset of 330 participants with valid identification was linked to the National Death Registry to determine the participants' survival status; they were followed up until 31 January 2020.

2.5. Statistical Analysis

All data analyses were performed using SAS statistical software (version 9.4; SAS Institute, Cary, NC, USA). Descriptive statistics are presented as the means $\pm$ standard deviations (SDs) and percentages for continuous and categorical variables, respectively. We evaluated the distribution of the differences within demographic and frailty severity by employing the one-way analysis of variance and the chi-square test or Fisher's exact test. We calculated the follow-up time from the date of the interview to either the date of their death or 31 January 2020. The Cox proportional hazards regression model was used to evaluate the association between frailty severity and all-cause mortality, and the additional analyses were in accordance with the DDS. However, the sample size was limited for frailty; thus, to ensure that there was statistical power, we combined the categories of prefrailty and frailty into "frailty status" in the joint effect analysis.

For all covariates except for body mass index (BMI), less than 6% of the values were missing and all of them were imputed to the mode value (all covariates were categorical variables). The proportion of missing values was high for BMI (34.5%). To avoid the massive imputation for a non-negligible number of participants, the exclusion of those with missing data, or the introduction of a bias, we included a missing class into the models for this variable.

3. Results

The baseline characteristics of the participants are shown by their frailty severity as shown in Table 1. A total of 330 participants (180 men [54.5%] and 150 women [45.5%]) were included in this study, and the proportions of prefrailty and frailty were 36.1% and 5.8%, respectively. The distribution of frailty severity was similar to that of the original dataset (Table S2). Frail older adults often had a lower education level, poorer perceived health and sleep quality, and a higher number of drug treatments. Those with prefrailty consumed a significantly higher proportion of dietary supplements. It is noteworthy that when they were compared to those features of men, frail, older women had a higher BMI, waist circumference, and hip circumference (all $p < 0.05$; Tables S3 and S4).

Table 1. Baseline characteristics by frailty severity among NAHSIT 2005–2008 older adults ($n = 330$).

	Frailty Severity			
	Robust	Prefrailty	Frailty	P-Trend
n, %	192 (58.2)	119 (36.1)	19 (5.8)	
Gender, %				0.763
Men	56.3	52.1	52.6	
Women	43.8	47.9	47.4	
Age (years), %	71.8 ± 5.47	73.5 ± 2.20	73.7 ± 5.44	0.008
65–69	42.7	30.3	36.8	0.234
70–74	25.0	26.9	15.8	
75–79	22.9	29.4	26.3	
≥80	9.38	13.5	21.1	
Education, %				0.048
Illiterate	19.8	26.9	36.8	
Primary school	60.9	45.4	42.1	
High school and above	19.3	27.7	21.1	
Smoking, %	35.4	36.2	36.8	0.985
Drink alcohol, %	42.1	37.1	36.8	0.659
Enough money, %				0.811
More than enough	8.47	10.3	5.88	
Just enough	56.6	50.0	52.9	
difficult	34.9	36.7	41.2	
Perceived health status, %				<0.001
Good	34.2	24.4	0.00	
Fair	54.6	42.6	17.7	
Poor	11.2	33.0	82.4	
Poor sleep quality, %	7.3	17.2	26.3	0.006
Number of diseases	2.03 ± 1.23	3.86 ± 2.20	5.84 ± 1.68	<0.001
Number of medicines	1.03 ± 1.03	2.06 ± 1.36	3.63 ± 1.95	<0.001
Cognitive impairment¶, %	15.3	22.4	31.6	0.342
Supplement use, %	43.3	62.2	36.8	0.002
BMI (kg/m^2)	24.4 ± 3.17	24.6 ± 3.52	26.7 ± 5.48	0.181
Waist circumference (cm)	85.3 ± 10.1	86.2 ± 10.2	91.5 ± 9.11	0.106
Hip circumference (cm)	92.6 ± 6.48	93.4 ± 6.68	96.8 ± 9.70	0.080
DDS (score)	4.74 ± 0.97	4.61 ± 1.10	4.32 ± 1.16	0.069

Abbreviations: NAHSIT, Nutrition and Health Survey in Taiwan; BMI, body mass index; DDS: dietary diversity score. Categorical variables are presented as a percentage, and continuous variables are presented as mean ± standard deviation (SD). General linear model regression and chi-square or Fisher's exact tests were used for continuous and categorical variables, respectively. Cognitive impairment¶ was defined by a short, portable mental status questionnaire (SPMSQ) with ≥3 errors.

The DDS for robust, prefrail, and frail participants were 4.74 ± 0.97, 4.61 ± 1.10, and 4.32 ± 1.16, respectively, but it did not show a significant difference for frailty severity in this study ($p = 0.069$). In addition, the mean consumption frequencies of egg, dairy products, noodles products, breakfast cereals, vegetables, mushroom, fresh fruit, nuts and seeds, poultry/livestock blood and other parts, and coffee increase with the DDS score. Those with a lower DDS had consumed less dairy, vegetables and fruits (Table S5).

During the median follow up of 11.7 years, 148 cases of death were recorded in the National Death Registry. Table 2 shows the associations between frailty severity and the risk of all-cause mortality. The crude model shows that frail, older people had a higher risk of mortality when it was compared with that of robust individuals (hazard ratio [HR]: 1.70, 95% confidence interval [CI]: 1.31–2.20). With adjustments for age and sex (Model 1), and additional adjustments for education level, smoking status, alcohol consumption, monetary status, sleep status, cognitive function, dietary supplement use, number of drugs that were being used (Model 2), DDS (Model 3), and perceived health status (Model 4), the HR was 3.31 (1.60–6.84).

Table 2. Association between frailty severity and risk of all-cause mortality in 2005–2008 NAHSIT older adults (n = 330).

	Frailty Severity			
	Robust	**Prefrailty**	**Frailty**	***p*-Value**
All-cause mortality				
Deceased/survival (*n*)	75/117	58/61	15/4	
Crude	1.00	1.42 (1.01–2.00)	3.73 (2.13–6.52)	<0.001
Model 1	1.00	1.30 (0.92–1.85)	3.80 (2.15–6.71)	<0.001
Model 2	1.00	1.42 (0.98–2.07)	5.09 (2.63–9.84)	<0.001
Model 3	1.00	1.42 (0.98–2.07)	5.10 (2.63–9.87)	<0.001
Model 4	1.00	1.19 (0.80–1.78)	3.31 (1.60–6.84)	0.004

The Cox proportional hazards model was estimated for hazard ratios. Model 1: adjusted for age and sex. Model 2: Model 1 plus education level, smoking status, alcohol consumption, monetary status, sleep status, cognitive function, supplement use, and the number of drug treatments that were being used. Model 3: Model 2 plus DDS. Model 4: Model 3 plus BMI.

We merged the categories of prefrailty and frailty into 'frailty status' in the analysis. The combined effects of frailty status and DDS (≤4 and >4) on all-cause mortality are presented in Table 3. In Model 1, older adults with a frailty status and a lower DDS had increased risks of mortality (HR: 2.01, 95% CI: 1.26–3.19) than those without frailty and a diverse diet. Further adjustments for Model 2 and Model 3 covariates did not modify these findings. Sensitivity analyses (adjusted for Model 1 covariates), excluding participants who died in the first year of the follow up, provided similar results (HR: 2.25, 95% CI: 1.41–3.59; p = 0.007 for mortality risk).

Table 3. Association between frailty status (combined prefrailty and frailty) stratified by DDS and all-cause mortality in 2005–2008 NAHSIT older adults (n = 329).

	Frailty Severity				
	Robust		**Prefrailty/Frailty**		
	DDS > 4	**DDS ≤ 4**	**DDS > 4**	**DDS ≤ 4**	***p*-Value**
All-cause mortality					
Deceased/survival	44/79	31/37	39/43	34/22	
Crude	1.00	1.27 (0.80–2.02)	1.50 (0.97–2.31)	2.25 (1.44–3.53)	0.004
Model 1	1.00	1.16 (0.72–1.85)	1.35 (0.88–2.09)	2.01 (1.26–3.19)	0.024
Model 2	1.00	1.05 (0.64–1.74)	1.56 (0.99–2.47)	1.79 (1.07–3.02)	0.062
Model 3	1.00	1.08 (0.54–2.19)	1.30 (0.72–3.74)	2.30 (1.11–4.75)	0.118

Cox proportional hazards model was estimated for hazard ratios. Model 1: adjusted for age and sex. Model 2: adjusted additionally for education level, smoking status, alcohol consumption, monetary status, sleep status, cognitive function, supplement use, and the number of drug treatments that were being used. Model 3: Model 2 plus BMI.

4. Discussion

In this prospective cohort study, we found that frailty which was based on the FRAIL scale definition can predict mortality among older Taiwanese people, and their survival may be improved by having dietary diversity. Older adults with a frailty status and a lower

DDS (DDS $\leq$ 4) had a higher risk of mortality when it was compared with that of robust adults or those with a frail status and a higher DDS (DDS > 4).

The systematic review and meta-analysis showed that the FRAIL scale can effectively identify frailty severity and predict disability among community-dwelling middle-aged and older people [23]. However, mortality studies on frailty and its severity are still limited [24]. We found that the FRAIL scale can predict the risk of mortality, and these findings did not change after adjustments were made for several covariates. In addition, the prevalence of prefrailty and frailty was 35.8% and 3.9%, respectively, among NAHSIT 2005–2008 participants (Table S2). Our previous study used representative data from the 2014–2016 NAHSIT of 1115 participants (aged $\geq$ 65 years), and found that the prevalence of prefrailty and frailty was 37.3% and 6.2%, respectively [11]. These results demonstrate that the prevalence of prefrailty and frailty is on the rise in Taiwan, and thus, it needs special attention.

We conducted the analysis by sex and found that frail, older women, but not men, had a higher BMI. These findings were consistent with a French longitudinal study of 1593 non-institutionalized older people that were aged $\geq$65 years. The results demonstrated that older women were approximately 2-fold more likely to be frail when they were obese, but there was no such correlation in men [25]. Moreover, a higher BMI was inversely associated with skeletal muscle mass, muscle strength, and bone mass, which may have been due to adipose tissue involvement in the complex bone-muscle interaction [26]. Therefore, older adults should maintain both a stable weight (BMI 24–26.9 kg/m^2) and also skeletal muscle mass to prevent frailty and subsequent disability and mortality [27,28].

In a Japanese prospective cohort study of 666 community-dwelling older adults, a "sugar and fat" dietary pattern had a positive association with frailty [29]. The consumption of a high sugar or fat diet has been shown to increase mitochondrial dysfunction and inflammation [30]. It may increase proteolysis and reduce skeletal muscle protein synthesis, and further reduce muscle function and strength [31]. Frailty may be the result of tissue damage that is caused by a poor antioxidant ability [32]. It has been suggested that frail, older adults tend to have a poor appetite or anorexia [33]. These diets were characterized by reducing the intake of certain food groups (e.g., protein-rich food and fruit and vegetables), and these adults often have a lower DDS, which typically infers the consumption of higher amounts of sugar and fat (e.g., refined foods and fatty meats) [14,33,34]. These aging-related overall dietary quality changes may increase the risk of frailty [16,35] and mortality [14,36].

The consumption of a single food or supplement alone cannot improve frailty [37]. Studies have been shown that dietary diversity is inversely associated with frailty [11,38]. Also, dietary interventions have been shown to improve frailty severity, which when they are combined with resistance training, lead to a great improvement in frailty severity [39]. Antioxidant polysaccharides from food sources such as vegetables, fruits, cereals, beans, mushrooms, tea, milk products, and shellfish are associated with antioxidant function [40]. Moreover, in this study, older adults with a higher DDS demonstrated a higher consumption frequency of vegetables, fruit, cereals, dairy products, and mushrooms (Table S5). Nevertheless, we did not find an association between dietary diversity and frailty severity in this cross-sectional study, but the DDS showed that there is a long-term protective effect. This study had some limitations. First, frailty and dietary data were collected at a baseline, which cannot assess the effects of the changes that occur or determine the causality of frailty. Second, we used one 24-h dietary recall test to measure the DDS, which may not completely capture the long-term DDS. Therefore, the composite responses from the FFQ were used to support the validity of the DDS from one 24-h dietary recall test in this study of the older adults. We have provided the distribution of food intake frequency as determined by the DDS in the Table S5. Moreover, these estimates, even though there is less variation in them, still illustrate our point, and the effect that they have may have been underestimated. Third, only 330 people with a valid identification could be linked to mortality. The information about mortality could be biased. The strength of this study is that it was a cohort study, with comprehensive socio-demographic characteristics as well as dietary and nutritional information.

5. Conclusions

In this prospective study, older adults with frailty had an increased risk of mortality. Higher dietary diversity might reduce the risk of mortality in older adults with prefrailty and frailty. Dietary diversity has health benefits. Health promotion interventions should emphasize the importance of eating a variety of foods, especially for older adults who tend to have reduced food intakes, and especially for those with a lower DDS who consume less dairy, vegetables and fruits. Further studies are needed to examine this association and investigate whether combined dietary and exercise interventions can reduce frailty and mortality in older adults.

Supplementary Materials: The following supporting information can be downloaded at: https://www.mdpi.com/article/10.3390/nu14183825/s1, Table S1: The definition of a serving for six food groups; Table S2: Baseline characteristics by frailty severity among NAHSIT 2005–2008 older adults (n = 968); Table S3: Baseline characteristics by frailty severity among NAHSIT 2005–2008 older men (n = 482); Table S4: Baseline characteristics by frailty severity among NAHSIT 2005–2008 older women (n = 486); Table S5: Distributions of food intake frequency by DDS among NAHSIT 2005–2008 older adults (n = 329).

Author Contributions: W.-C.H.: Study conception and design, data analysis and interpretation, and manuscript drafting and revision. Y.-C.H., M.-S.L., J.-Y.D. and W.-H.P.: Study conception and design, and manuscript revision. H.-Y.C.: Research coordination; study conception and design; data acquisition, analysis, and interpretation; and manuscript drafting and revision. All authors have read and agreed to the published version of the manuscript.

Funding: Funding for this study was provided by grants from the Health Promotion Administration, Ministry of Health and Welfare in Taiwan (DOH94-fs-6-4), and a grant from Academia Sinica. These funding sources had no role in the design, methods, subject recruitment, data collection, analysis, or preparation of this manuscript.

Institutional Review Board Statement: This study was approved by the Research Ethics Committee of the National Health Research Institute (EC1080701-E; Miaoli, Taiwan) on 30 July 2019.

Informed Consent Statement: Informed consent was obtained from all subjects involved in the study.

Data Availability Statement: Restrictions apply to the availability of these data. Data was obtained from Wen-Harn Pan, and are available Wen-Harn Pan with the permission of https://www.ibms.sinica.edu.tw/wen-harn-pan/ (accessed on 31 July 2022).

Acknowledgments: Data analyzed in this paper (article) were collected by the research project "Nutrition and Health Survey in Taiwan 2005–2008" sponsored by the Department of Health in Taiwan. This research project was carried out by the Center for Survey Research of the Research Center for Humanities and Social Sciences of Academia Sinica, directed by Wen-Harn Pan. The Office of Nutrition Survey, the Institute of Biomedical Sciences, Academia Sinica is responsible for data distribution. The assistance provided by the institutes and efforts made by all contributed to the survey are greatly appreciated. The views expressed herein are solely those of the authors. Funding sources had no role in the design, methods, subject recruitment, data collection, analysis, or preparation of this manuscript.

Conflicts of Interest: The authors declare no conflict of interest.

References

1. Clegg, A.; Young, J.; Iliffe, S.; Rikkert, M.O.; Rockwood, K. Frailty in elderly people. *Lancet* **2013**, *381*, 752–762. [CrossRef]
2. Fried, L.P.; Tangen, C.M.; Walston, J.; Newman, A.B.; Hirsch, C.; Gottdiener, J.; Seeman, T.; Tracy, R.; Kop, W.J.; Burke, G.; et al. Frailty in older adults: Evidence for a phenotype. *J. Gerontol. A Biol. Sci. Med. Sci.* **2001**, *56*, M146–M156. [CrossRef] [PubMed]
3. Siriwardhana, D.D.; Hardoon, S.; Rait, G.; Weerasinghe, M.C.; Walters, K.R. Prevalence of frailty and prefrailty among community-dwelling older adults in low-income and middle-income countries: A systematic review and meta-analysis. *BMJ Open* **2018**, *8*, e018195. [CrossRef] [PubMed]

4. Morley, J.E.; Vellas, B.; van Kan, G.A.; Anker, S.D.; Bauer, J.M.; Bernabei, R.; Cesari, M.; Chumlea, W.C.; Doehner, W.; Evans, J.; et al. Frailty consensus: A call to action. *J. Am. Med. Dir. Assoc.* **2013**, *14*, 392–397. [CrossRef]
5. World Health Organization. Global Action Plan for the Prevention and Control of NCDs 2013–2020. Available online: http://www.who.int/nmh/events/ncd_action_plan/en/ (accessed on 31 July 2022).
6. Gopinath, B.; Russell, J.; Flood, V.M.; Burlutsky, G.; Mitchell, P. Adherence to dietary guidelines positively affects quality of life and functional status of older adults. *J. Acad. Nutr. Diet.* **2014**, *114*, 220–229. [CrossRef]
7. Stewart, R.A.H. Primary Prevention of Cardiovascular Disease with a Mediterranean Diet Supplemented with Extra-Virgin Olive Oil or Nuts. *N. Engl. J. Med.* **2018**, *379*, 1388.
8. Chuang, S.Y.; Lo, Y.L.; Wu, S.Y.; Wang, P.N.; Pan, W.H. Dietary Patterns and Foods Associated with Cognitive Function in Taiwanese Older Adults: The Cross-sectional and Longitudinal Studies. *J. Am. Med. Dir. Assoc.* **2019**, *20*, 544–550.e544. [CrossRef]
9. Lee, M.S.; Huang, Y.C.; Su, H.H.; Lee, M.Z.; Wahlqvist, M.L. A simple food quality index predicts mortality in elderly Taiwanese. *J. Nutr. Health Aging* **2011**, *15*, 815–821. [CrossRef]
10. Kant, A.K.; Schatzkin, A.; Harris, T.B.; Ziegler, R.G.; Block, G. Dietary diversity and subsequent mortality in the First National Health and Nutrition Examination Survey Epidemiologic Follow-up Study. *Am. J. Clin. Nutr.* **1993**, *57*, 434–440. [CrossRef]
11. Huang, W.C.; Huang, Y.C.; Lee, M.S.; Chang, H.Y.; Doong, J.Y. Frailty Severity and Cognitive Impairment Associated with Dietary Diversity in Older Adults in Taiwan. *Nutrients* **2021**, *13*, 418. [CrossRef]
12. Lo, Y.L.; Hsieh, Y.T.; Hsu, L.L.; Chuang, S.Y.; Chang, H.Y.; Hsu, C.C.; Chen, C.Y.; Pan, W.H. Dietary Pattern Associated with Frailty: Results from Nutrition and Health Survey in Taiwan. *J. Am. Geriatr. Soc.* **2017**, *65*, 2009–2015. [CrossRef]
13. Chen, R.C.; Chang, Y.H.; Lee, M.S.; Wahlqvist, M.L. Dietary quality may enhance survival related to cognitive impairment in Taiwanese elderly. *Food Nutr. Res.* **2011**, *55*, 7387. [CrossRef]
14. Huang, Y.C.; Wahlqvist, M.L.; Lee, M.S. Appetite predicts mortality in free-living older adults in association with dietary diversity. A NAHSIT cohort study. *Appetite* **2014**, *83*, 89–96. [CrossRef]
15. Wahlqvist, M.L.; Xiu, L.; Lee, M.S.; Chen, R.C.; Chen, K.J.; Li, D. Dietary diversity no longer offsets the mortality risk of hyperhomocysteinaemia in older adults with diabetes: A prospective cohort study. *Asia Pac. J. Clin. Nutr.* **2016**, *25*, 414–423.
16. Hengeveld, L.M.; Wijnhoven, H.A.H.; Olthof, M.R.; Brouwer, I.A.; Simonsick, E.M.; Kritchevsky, S.B.; Houston, D.K.; Newman, A.B.; Visser, M. Prospective Associations of Diet Quality with Incident Frailty in Older Adults: The Health, Aging, and Body Composition Study. *J. Am. Geriatr. Soc.* **2019**, *67*, 1835–1842. [CrossRef]
17. Berry, S.D.; Ngo, L.; Samelson, E.J.; Kiel, D.P. Competing risk of death: An important consideration in studies of older adults. *J. Am. Geriatr. Soc.* **2010**, *58*, 783–787. [CrossRef]
18. Tu, S.H.; Chen, C.; Hsieh, Y.T.; Chang, H.Y.; Yeh, C.J.; Lin, Y.C.; Pan, W.H. Design and sample characteristics of the 2005–2008 Nutrition and Health Survey in Taiwan. *Asia Pac. J. Clin. Nutr.* **2011**, *20*, 225–237.
19. Abellan van Kan, G.; Rolland, Y.M.; Morley, J.E.; Vellas, B. Frailty: Toward a clinical definition. *J. Am. Med. Dir. Assoc.* **2008**, *9*, 71–72. [CrossRef]
20. Woo, J.; Leung, J.; Morley, J.E. Comparison of frailty indicators based on clinical phenotype and the multiple deficit approach in predicting mortality and physical limitation. *J. Am. Geriatr. Soc.* **2012**, *60*, 1478–1486. [CrossRef]
21. Chen, J.Y.; Wu, S.J.; Tzeng, M.S.; Kao, M.D. Nutrition and health survey in Taiwan (NAHSIT) 1993-1996: Dietary nutrient intakes assessed by 24-h recall. *Nutr. Sci. J.* **1999**, *24*, 11–39.
22. Huang, Y.C.; Lee, M.S.; Pan, W.H.; Wahlqvist, M.L. Validation of a simplified food frequency questionnaire as used in the Nutrition and Health Survey in Taiwan (NAHSIT) for the elderly. *Asia Pac. J. Clin. Nutr.* **2011**, *20*, 134–140.
23. Kojima, G. Quick and Simple FRAIL Scale Predicts Incident Activities of Daily Living (ADL) and Instrumental ADL (IADL) Disabilities: A Systematic Review and Meta-analysis. *J. Am. Med. Dir. Assoc.* **2018**, *19*, 1063–1068. [CrossRef]
24. Kojima, G. Frailty Defined by FRAIL Scale as a Predictor of Mortality: A Systematic Review and Meta-analysis. *J. Am. Med. Dir. Assoc.* **2018**, *19*, 480–483. [CrossRef]
25. Monteil, D.; Walrand, S.; Vannier-Nitenberg, C.; Van Oost, B.; Bonnefoy, M. The Relationship between Frailty, Obesity and Social Deprivation in Non-Institutionalized Elderly People. *J. Nutr. Health Aging* **2020**, *24*, 821–826. [CrossRef]
26. Choi, K.M. Sarcopenia and sarcopenic obesity. *Korean J. Intern. Med.* **2016**, *31*, 1054–1060. [CrossRef]
27. Chuang, S.Y.; Chang, H.Y.; Lee, M.S.; Chia-Yu Chen, R.; Pan, W.H. Skeletal muscle mass and risk of death in an elderly population. *Nutr. Metab. Cardiovasc. Dis.* **2014**, *24*, 784–791. [CrossRef]
28. Ishii, H.; Tsutsumimoto, K.; Doi, T.; Nakakubo, S.; Kim, M.; Kurita, S.; Shimada, H. Effects of comorbid physical frailty and low muscle mass on incident disability in community-dwelling older adults: A 24-month follow-up longitudinal study. *Maturitas* **2020**, *139*, 57–63. [CrossRef]
29. Huang, C.H.; Martins, B.A.; Okada, K.; Matsushita, E.; Uno, C.; Satake, S.; Kuzuya, M. A 3-year prospective cohort study of dietary patterns and frailty risk among community-dwelling older adults. *Clin. Nutr.* **2021**, *40*, 229–236. [CrossRef]
30. Valencia, A.P.; Nagaraj, N.; Osman, D.H.; Rabinovitch, P.S.; Marcinek, D.J. Are fat and sugar just as detrimental in old age? *GeroScience* **2021**, *43*, 1615–1625. [CrossRef]
31. Liguori, I.; Russo, G.; Curcio, F.; Bulli, G.; Aran, L.; Della-Morte, D.; Gargiulo, G.; Testa, G.; Cacciatore, F.; Bonaduce, D.; et al. Oxidative stress, aging, and diseases. *Clin. Interv. Aging* **2018**, *13*, 757–772. [CrossRef]
32. Soysal, P.; Stubbs, B.; Lucato, P.; Luchini, C.; Solmi, M.; Peluso, R.; Sergi, G.; Isik, A.T.; Manzato, E.; Maggi, S.; et al. Inflammation and frailty in the elderly: A systematic review and meta-analysis. *Ageing Res. Rev.* **2016**, *31*, 1–8. [CrossRef] [PubMed]

33. Donini, L.M.; Poggiogalle, E.; Piredda, M.; Pinto, A.; Barbagallo, M.; Cucinotta, D.; Sergi, G. Anorexia and eating patterns in the elderly. *PLoS ONE.* **2013**, *8*, e63539. [CrossRef] [PubMed]
34. Vitolins, M.Z.; Quandt, S.A.; Bell, R.A.; Arcury, T.A.; Case, L.D. Quality of diets consumed by older rural adults. *J. Rural Health* **2002**, *18*, 49–56. [CrossRef] [PubMed]
35. Tsutsumimoto, K.; Doi, T.; Makizako, H.; Hotta, R.; Nakakubo, S.; Makino, K.; Suzuki, T.; Shimada, H. Aging-related anorexia and its association with disability and frailty. *J. Cachexia Sarcopenia Muscle* **2018**, *9*, 834–843. [CrossRef] [PubMed]
36. Lee, M.S.; Huang, Y.C.; Wahlqvist, M.L. Chewing ability in conjunction with food intake and energy status in later life affects survival in Taiwanese with the metabolic syndrome. *J. Am. Geriatr. Soc.* **2010**, *58*, 1072–1080. [CrossRef] [PubMed]
37. Dominguez, L.J.; Barbagallo, M. The relevance of nutrition for the concept of cognitive frailty. *Curr. Opin. Clin. Nutr. Metab. Care* **2017**, *20*, 61–68. [CrossRef]
38. Motokawa, K.; Watanabe, Y.; Edahiro, A.; Shirobe, M.; Murakami, M.; Kera, T.; Kawai, H.; Obuchi, S.; Fujiwara, Y.; Ihara, K.; et al. Frailty Severity and Dietary Variety in Japanese Older Persons: A Cross-Sectional Study. *J. Nutr. Health Aging* **2018**, *22*, 451–456. [CrossRef]
39. Hsieh, T.J.; Su, S.C.; Chen, C.W.; Kang, Y.W.; Hu, M.H.; Hsu, L.L.; Wu, S.Y.; Chen, L.; Chang, H.Y.; Chuang, S.Y.; et al. Individualized home-based exercise and nutrition interventions improve frailty in older adults: A randomized controlled trial. *Int. J. Behav. Nutr. Phys. Act.* **2019**, *16*, 119.
40. Li, H.; Ding, F.; Xiao, L.; Shi, R.; Wang, H.; Han, W.; Huang, Z. Food-Derived Antioxidant Polysaccharides and Their Pharmacological Potential in Neurodegenerative Diseases. *Nutrients* **2017**, *9*, 778. [CrossRef]

nutrients

MDPI

Article

Nutrition Behavior and Physical Activity of Middle-Aged and Older Adults in Saudi Arabia

Rouba Khalil Naaman

Clinical Nutrition Department, Faculty of Applied Medical Sciences, King Abdulaziz University, P.O. Box 80215, Jeddah 21589, Saudi Arabia; rnaaman@kau.edu.sa

Abstract: As people get older, their nutritional status deteriorates, resulting in increased vulnerability to chronic diseases. The adoption of a healthy lifestyle has been linked to improved health throughout the aging process. The current study aimed to assess nutritional behaviors, dietary patterns, and physical activity among middle-aged and older adults in Saudi Arabia. An electronic questionnaire was completed between September and November 2021 by 419 participants aged 45 years and older. Of those, 65% reported that nutrition was important to them and 19% stated that they were consuming a healthy diet. Participants reported consuming an average of around 6 servings/week each of fruit and vegetables, with mean intake scores of 5.92 ± 0.25 and 5.57 ± 0.22, respectively. It was reported that around 3 servings/week of red meat, 4 servings/week of poultry, and 1 serving/week of fish were consumed, with mean intake scores of 2.65 ± 0.13, 4.34 ± 0.16, and 1.36 ± 0.08, respectively. Most of the participants (60%) reported being inactive. Middle-aged and older adults living in Saudi Arabia have poor dietary patterns and nutritional behaviors. Education and guidance on nutrition are needed for this population to help them improve their diet and lifestyle.

Keywords: older adults; middle-aged; nutrition behavior; dietary pattern; physical activity

1. Introduction

The percentage of the world's population that is elderly is growing rapidly, and it is expected that there will be 1.4 billion adults aged 60 years and over by 2050 [1]. According to the United Nations 2021, in Saudi Arabia, the percentage of the country's population that is elderly increased from 4.5% in 2010 to 5.9% in 2020, and this is estimated to further increase in the coming years [2]. This ageing of the population comes with challenges that need to be considered by governments and health services, including an increased risk of chronic illnesses and co-morbidities, and consequent increased healthcare costs.

As people get older, their nutritional status deteriorates as a result of several physiological and social age-related changes. Deteriorated digestive function, loss of muscle mass, impaired functional status, loss of appetite, poor oral health, depression, social isolation, and poor economic status are some examples of changes that often accompany aging, which, in turn, result in poor health outcomes such as increased vulnerability to chronic diseases, infections, physical disability, and cognitive impairments [1,3,4]. Healthy aging refers to the maintenance of physical and cognitive health with a reduced risk of chronic diseases with aging [5]. The link between a healthy lifestyle and healthy aging has recently been extensively studied [6,7]. The adoption of healthy habits such as the consumption of a balanced diet and engagement in physical activity has been shown to be associated with improved overall health and functional capacity in old age [8].

The quality of one's diet is one of the key factors that can be modified to prevent the development of chronic diseases with aging. A previous study that assessed the intake of Saudi adults aged 15 and older showed that only a small percentage of the Saudi population met the Saudi dietary guidelines [9]. However, there is still a paucity of studies that assess the dietary behavior of middle-aged and older adults' in Saudi Arabia. Therefore, to help fill this gap in the research, the current study aims to assess nutritional behaviors, dietary patterns, and physical activity in middle-aged and older adults in Saudi Arabia.

2. Materials and Methods

2.1. Study Design

This cross-sectional study was approved by the Unit of the Biomedical Ethics Research Committee at King Abdulaziz University (Jeddah, Saudi Arabia) (reference no. 426-21). A convenience sample of 419 participants was recruited electronically and asked to complete an online questionnaire. All study participants gave their consent to participate in this study at the beginning of the online questionnaire.

2.2. Participants and Recruitment

The inclusion criteria were being a male or female citizen or resident of Saudi Arabia aged ≥45 years or older. Between September and November 2021, the study participants completed an electronic questionnaire. The questionnaire was created via Microsoft Forms, version Microsoft 365 (Microsoft Corporation, Redmond, WA, USA) and distributed on WhatsApp (Facebook, Inc., Menlo Park, CA, USA) and Twitter (Twitter, Inc., San Francisco, CA, USA). WhatsApp was used to share the questionnaire link with the authors' relatives and friends so that they could participate in the study and share the questionnaire link with their own contacts. The study's information and participation link were also posted on Twitter, and to reach a wider group of people from various regions of Saudi Arabia, the link to the questionnaire was promoted.

2.3. Questionnaire

This online questionnaire was intended to assess dietary and nutritional behaviors and physical activity in middle-aged and older adults in Saudi Arabia. To improve the quality of the questionnaire, three nutrition experts (MSc and PhD holders) reviewed the questionnaire in terms of relevance, clarity, simplicity, and ambiguity, and their comments on it were taken into account. Modifications and changes that were made included correcting linguistic and grammatical errors, rewording, and adding options to some questions regarding participants' sociodemographic characteristics and dietary patterns. For pre-testing, the revised questionnaire was shared with four persons aged 45 and over (who were not included in the main study) so that they could check the understandability of the questions and answers, and they could assess the average time needed for completion. Changes were then made based on the pre-testing's feedback, including revising and reformulating some questions and their options. As a result of these amendments, the final version of the questionnaire, which needed around 15 min to complete, consisted of four main sections. It was created and distributed in Arabic. Participants were provided with information regarding the study at the beginning of the questionnaire, including the study's main aims and inclusion criteria, the confidentiality of the data being collected, and the expected time needed to complete the questionnaire.

In section one, participants' socio-demographic data were collected, including age, gender, nationality, marital status, city of residence, work status, educational level, monthly income, living status, presence of chronic disease, and smoking status. Participants' anthropometric measurements, including self-reported weight in kilograms and height in centimeters, were also collected. Both measurements were then used to determine their body mass index (BMI).

In section two, participants were asked about how important nutrition was to them and to rate the healthfulness of their diet by selecting one of the following options: important/healthy, somewhat important/somewhat healthy, or not important/not healthy. Questions about whether they received any dietary advice or information related to dietary intake and the source of the advice were also included. Regarding nutrition facts labels, participants were asked about how frequently they read such labels when buying new food products and, if they indicated that they did not read them, why not.

In section three, participants were asked about their consumption of certain food groups, including starch, fruits, vegetables, milk and dairy products, red meat, poultry, fish, and legumes. For each food group, participants indicated the frequency of their consumption of the group by choosing one of the following options: I do not eat it at all, 1–3 times/month, 1 time/week, 2 times/week, 3 times/week, 4 times/week, 5 times/week, 6 times/week, 1 time/day, or 2 or more times/day. Second, they were asked about the estimated portion size they would eat each time they consumed a given food group and were given the following answers to choose from: I do not eat it at all, <1 portion, 1 portion, 2 portions, 3–4 portions, or 5 or more portions. For each food group, participants were provided with examples of food items and the estimated size of one portion (e.g., starch: ½ cup cereals or 1 slice bread or 1 potato; fruits: 1 fruit or ½ cup juice or dried fruits; vegetables: 1 cup fresh or ½ cup cooked; milk and dairy products: 1 glass or cup or 1 slice of cheese; red meat/poultry/fish: 30 g; and legumes: 1 cup). To enable a better understanding of the portion sizes, pictures of the size of one portion were given for each food group.

In section four, participants were asked whether they performed physical activity. They were also asked about the frequency and type of physical activity they performed per week, and those who did not perform any type of physical activity were asked why not.

2.4. Data Analysis

The frequency of intake of each food group and the portion size consumed each time a given food group was consumed were all converted to the number of servings consumed per week. The frequency of the food groups' intake data was coded as follows: I do not eat it at all = 0, 1–3 times/month = 0.5, 1 time/week = 1, 2 times/week = 2, 3 times/week = 3, 4 times/week = 4, 5 times/week = 5, 6 times/week = 6, 1 time/day = 7, or 2 or more times/day = 14. Data about the portion size consumed each time of consumption were coded as the following: I do not eat it at all = 0, <1 portion = 0.5, 1 portion = 1, 2 portions = 2, 3–4 portions = 3, or 5 or more portions = 5. To obtain the number of servings consumed each week, the scores of frequency were multiplied by the scores of the consumed portion sizes. Then, the consumption of each food group was rated on a 7-point scale. A score of 0 was given when no consumption was reported, a score of 0.5 when <1 serving/week was reported, a score of 1 when 1 serving/week was reported, 2 for 2–3 servings/week, 4 for 4–6 servings/week, 7 for 7 servings/weeks or 1 serving/day, and a score of 14 when ≥14 servings/week or more or ≥2 servings/day were reported.

2.5. Sample Size Calculation

The study sample size was computed by using Raosoft® software, version 2004 (www.raosoft.com/samplesize.html) (Raosoft, Inc., Seattle, WA, USA) (accessed on 12 November 2021). The required sample size was calculated based on the population living in Saudi Arabia (aged ≥45 years old) as reported by the Saudi General Authority for Statistics in 2020 [10]. The anticipated frequency is 50%, with a 95% confidence level, a 5% margin error, and a design effect of 1; therefore, a minimum sample of 385 participants was required to be enrolled.

2.6. Statistical Analysis

A statistical analysis was performed using Minitab® statistical software, version 19 (Penn State University, State College, PA, USA). The Anderson–Darling test was used to evaluate the distribution of the variables. Categorical data were expressed as a number and a percentage, and continuous data were expressed as mean and standard deviation. Differences between categorical variables were assessed with the Chi-square test. The Kruskal–Wallis test was used to test the differences between continuous variables. A p value of <0.05 was considered to be statistically significant.

3. Results

3.1. Characteristics of the Study Participants

This study's questionnaire was completed by 419 participants. Table 1 presents the general characteristics of the population studied. Forty-six percent of the participants were aged 45–54 years old, 66% were males, 91% were Saudis, and 63% of them were from the Western Region of Saudi Arabia. The majority of the participants were married (86%), lived with others (62%), and received university-level education (61%). Forty-four percent of the recruited participants were retired, while 38% of them were employed. Most of them had monthly incomes higher than 10,000 Saudi riyals (66%).

Table 1. General characteristics of the study participants (n = 419) [1].

Variables	N	%
Age (years)		
45–54	194	46
55–64	169	40
≥65	56	14
Mean ± SD	55.5 ± 7	
Gender		
Male	278	66
Female	141	34
Nationality		
Saudi	382	91
Non-Saudi [2]	37	9
Region		
Western Region	262	63
Central Region	98	23
Eastern Region	33	8
Northern Region	9	2
Southern Region	17	4
Marital status		
Single	24	6
Married	362	86
Divorced	21	5
Widower	12	3
Living situation		
Living alone	158	38
Living with others	261	62
Education level		
High school education or less	81	19
University education	253	61
Higher education	85	20
Work status		
Employed	161	38
Freelance job	27	7
Retired	184	44
Unemployed	47	11

Table 1. *Cont.*

Variables	N	%
Income (SR)		
<2000	37	9
2000–5000	34	8
5000–7000	23	6
7000–10,000	48	11
>10,000	277	66
Main medical diagnosis		
No diseases	153	37
Heart diseases	144	34
Respiratory disorders	25	6
Gastrointestinal diseases	29	7
Cancer	10	3
Renal disorders	9	2
Liver disorders	4	1
Diabetes	102	24
Iron deficiency anemia	19	4
Osteoporosis	29	7
Others [3]	23	5
Smoking		
Yes	93	22
No	257	61
Ex-smoker	69	17
BMI category [4]		
Underweight	4	1
Normal	84	20
Overweight	171	41
Obese	160	38

[1] Data presented as number and percentage. [2] Non-Saudis including Egyptian, Palestinian, Yemeni, Sudanese, and Jordanian. [3] Other diseases including hypothyroidism, gout, and multiple sclerosis. [4] Self-reported weight and height used to calculate the BMI. The BMI categories are underweight (<18.5 kg/m^2), normal weight (18.5–24.9 kg/m^2), overweight (25.0–29.9 kg/m^2), and obese (≥30 kg/m^2). Abbreviations: SD, standard deviation; SR, Saudi Riyals; BMI, body mass index.

Regarding the participants' main medical diagnoses, 37% reported that they were not suffering from any medical disorder, and 34% reported that they had heart diseases. Most of the participants were nonsmokers (61%). Forty-one percent of them reported being overweight and thirty-eight percent being obese.

3.2. Dietary Behavior Self-Evaluation

The majority of the study participants (65%) reported that nutrition was important to them. A significant difference was seen with monthly income ($p < 0.05$). However, no differences were found with other variables. Regarding participants' self-evaluation of the healthfulness of their diet, most of them (66%) stated that their diet was somewhat healthy and 19% stated that they were consuming a healthy diet. Significant differences were seen with age and monthly income ($p < 0.05$), whereas other variables such as gender, marital status, living situation, education, and work status showed no associations. The participants' nutrition behavior self-assessment is presented in Table 2.

Sixty percent of the study participants reported that they had received dietary advice or information related to their dietary intake previously, with half of them (50%) stating that the source of this advice was from a health specialist, including dietitians and physicians. A quarter (25%) of those who reported that they had previously received dietary advice indicated that the source of the advice was from social media platforms such as Instagram, Twitter, Snapchat, and Facebook (Figure 1).

Table 2. Nutrition behavior self-assessment (n = 419) [1].

Variables	How Important is Nutrition for You?			How Would You Rate the Healthfulness of Your Overall Diet?		
	Important	Somewhat Important	Not Important	Healthy	Somewhat Healthy	Not Healthy
Total (%)	65	30	5	19	66	15
Age (years)						
45–54	122 (29)	59 (14)	13 (3)	30 (7)	125 (30)	39 (9)
55–64	110 (26)	52 (12)	7 (2)	38 (9)	116 (28)	15 (4)
≥65	40 (10)	15 (4)	1 (0)	14 (3)	34 (8)	8 (2)
p-value	0.517			0.022		
Gender						
Male	184 (44)	84 (20)	10 (2)	56 (13)	180 (43)	42 (10)
Female	88 (21)	42 (10)	11 (3)	26 (6)	95 (23)	20 (5)
p-value	0.173			0.865		
Marital status						
Single	14 (3)	8 (2)	2 (0)	5 (1)	13 (3)	6 (1)
Married	239 (57)	108 (26)	15 (4)	72 (17)	240 (57)	50 (12)
Divorced	9 (2)	9 (2)	3 (1)	5 (1)	12 (3)	4 (1)
Widower	10 (3)	1 (0)	1 (0)	0 (0)	10 (3)	2 (1)
p-value	0.090			0.412		
Living situation						
Living alone	101 (24)	52 (12)	5 (1)	29 (7)	110 (26)	19 (5)
Living with others	171 (41)	74 (18)	16 (4)	53 (12)	165 (40)	43 (10)
p-value	0.294			0.346		
Education						
High school or less	51 (12)	24 (6)	6 (3)	18 (4)	51 (12)	12 (3)
University education	160 (38)	79 (19)	14 (2)	46 (11)	168 (40)	39 (9)
Higher education	61 (15)	23 (5)	1 (0)	18 (4)	56 (14)	11 (3)
p-value	0.319			0.906		
Work status						
Employed	107 (26)	45 (11)	9 (2)	27 (6)	105 (25)	29 (7)
Freelance Job	18 (4)	9 (2)	0 (0)	6 (2)	17 (4)	4 (1)
Retired	120 (29)	56 (13)	8 (2)	39 (9)	122 (29)	23 (5)
Unemployed	27 (6)	16 (4)	4 (1)	10 (2)	31 (8)	6 (2)
p-value	0.698			0.806		
Income (SR)						
<2000	19 (5)	14 (3)	4 (1)	10 (2)	22 (5)	5 (1)
2000–5000	22 (5)	8 (2)	4 (1)	7 (1)	20 (5)	7 (2)
5000–7000	14 (3)	8 (2)	1 (1)	3 (1)	13 (3)	7 (2)
7000–10,000	26 (6)	16 (4)	6 (1)	4 (1)	32 (8)	12 (3)
>10,000	191 (46)	80 (19)	6 (1)	58 (14)	188 (45)	31 (7)
p-value	0.011			0.038		
BMI category [2]						
Underweight	2 (0)	2 (0)	0 (0)	0 (0)	2 (0)	2 (0)
Normal weight	53 (13)	26 (6)	5 (1)	23 (5)	54 (13)	7 (2)
Overweight	110 (26)	53 (13)	8 (2)	35 (8)	115 (28)	21 (5)
Obese	107 (26)	45 (11)	8 (2)	24 (6)	104 (25)	32 (8)
p-value	0.959			0.068		

[1] Data presented as number and percentage. [2] Self-reported weight and height used to calculate the BMI. The BMI categories are underweight (<18.5 kg/m^2), normal weight (18.5–24.9 kg/m^2), overweight (25.0–29.9 kg/m^2), and obese (≥30 kg/m^2). Differences between the three groups were assessed via Chi-square test. Abbreviations: SR, Saudi Riyals; BMI, body mass index.

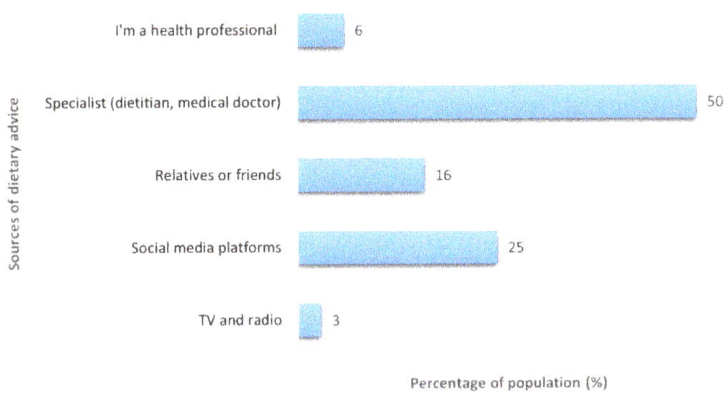

Figure 1. Participants' sources of dietary advice (n = 252).

Regarding reading nutrition facts labels when buying new food products, around half of the participants (51%) stated that they did not read the labels. Unclear fonts and designs and a lack of interest in knowing the nutritional content of food products were two of the most common reasons reported for not reading the nutrition facts label (Figure 2). For those who stated that they read the labels (49%), 53% of them said that they read it most of the time, 29% that they sometimes read it, and only 18% that they always read the labels.

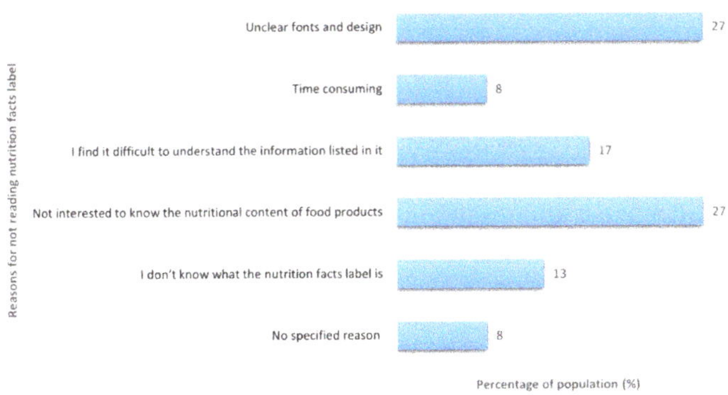

Figure 2. Reasons for not reading nutrition facts labels (n = 419).

3.3. Dietary Intake Assessment

Table 3 shows the scores of the weekly intake of different food groups based on the frequency and quantity consumed. Across all study participants, the mean score for starch intake per week was 8.09 ± 0.24. There were no variables associated with starch intake ($p > 0.05$). With regard to fruits, the total mean score of weekly intake was 5.92 ± 0.25. Age, living situation, work status, and income were significantly associated with fruit consumption ($p < 0.05$). The mean score of vegetable intake was 5.57 ± 0.22, with age and gender being the variables associated with vegetable intake ($p < 0.05$). The total mean score of weekly intake of milk and dairy product was 5.56 ± 0.23. There were no variables associated with intake of milk and dairy products ($p > 0.05$). In terms of intake of red meat and poultry, the mean weekly intake scores were 2.65 ± 0.13 and 4.34 ± 0.16, respectively. Gender, marital status, work status, and income were variables associated with red meat and poultry intakes ($p < 0.05$). The mean score for weekly fish intake was 1.36 ± 0.08. Age, gender, marital status, work status, and income were all significantly associated with fish intake ($p < 0.01$). With regard to legumes, the total mean score of the intake per week was 2.03 ± 0.12. Gender and income were the variables that were found to be significantly associated with legume intake ($p < 0.05$).

Table 3. Dietary intake scoring of different food groups in servings per week ($n = 419$) [1].

Variables	Starch	Fruits	Vegetables	Milk and Dairy Products	Red Meat	Poultry	Fish	Legumes
Mean ± SD	8.09 ± 0.24	5.92 ± 0.25	5.57 ± 0.22	5.56 ± 0.23	2.65 ± 0.13	4.34 ± 0.16	1.36 ± 0.08	2.03 ± 0.12
Age (years)								
45–54	7.92 ± 0.35	4.28 ± 0.33	4.77 ± 0.30	5.74 ± 0.36	2.59 ± 0.22	4.79 ± 0.26	1.10 ± 0.12	2.09 ± 0.21
55–64	8.02 ± 0.37	7.25 ± 0.41	6.44 ± 0.37	5.43 ± 0.35	2.57 ± 0.18	3.96 ± 0.24	1.58 ± 0.13	1.89 ± 0.14
≥65	8.90 ± 0.67	7.61 ± 0.70	5.75 ± 0.60	5.32 ± 0.55	3.07 ± 0.43	3.95 ± 0.39	1.61 ± 0.18	2.23 ± 0.30
p-value	0.817	<0.001	0.027	0.542	0.267	0.089	<0.001	0.143
Gender								
Male	8.14 ± 0.29	5.52 ± 0.30	5.20 ± 0.26	5.40 ± 0.27	2.94 ± 0.17	4.78 ± 0.21	1.54 ± 0.10	2.25 ± 0.15
Female	7.99 ± 0.40	6.12 ± 0.46	6.30 ± 0.41	5.88 ± 0.43	2.06 ± 0.22	3.48 ± 0.26	1.02 ± 0.12	1.59 ± 0.17
p-value	0.903	0.948	0.035	0.649	<0.001	<0.001	<0.001	<0.001
Marital status								
Single	8.06 ± 1.20	3.54 ± 0.81	5.22 ± 0.93	4.14 ± 0.90	1.91 ± 0.45	2.68 ± 0.41	1.35 ± 0.35	1.56 ± 0.31
Married	8.13 ± 0.25	6.08 ± 0.27	5.65 ± 0.24	5.56 ± 0.25	2.76 ± 0.15	4.57 ± 0.18	1.42 ± 0.09	2.10 ± 0.13
Divorced	6.97 ± 0.95	6.12 ± 1.22	4.98 ± 1.08	6.85 ± 0.88	2.59 ± 0.71	3.21 ± 0.72	0.83 ± 0.21	1.50 ± 0.31
Widower	9.00 ± 1.35	5.67 ± 1.60	4.96 ± 1.38	6.29 ± 1.78	0.87 ± 0.17	2.75 ± 0.67	0.41 ± 0.10	1.83 ± 0.60
p-value	0.698	0.112	0.655	0.094	0.009	0.002	0.002	0.633
Living situation								
Living alone	8.31 ± 0.38	6.51 ± 0.42	5.78 ± 0.37	5.49 ± 0.37	2.74 ± 0.24	4.25 ± 0.25	1.41 ± 0.12	1.87 ± 0.14
Living with others	7.95 ± 0.31	5.56 ± 0.32	5.45 ± 0.28	5.60 ± 0.29	2.59 ± 0.16	4.40 ± 0.22	1.33 ± 0.11	2.13 ± 0.17
p-value	0.423	0.045	0.504	0.888	0.982	0.808	0.194	0.599
Education								
High school or less	7.09 ± 0.50	5.53 ± 0.58	5.74 ± 0.50	5.90 ± 0.58	2.45 ± 0.30	4.63 ± 0.42	1.01 ± 0.10	2.09 ± 0.28
University education	8.13 ± 0.31	5.86 ± 0.33	5.26 ± 0.28	5.55 ± 0.29	2.66 ± 0.18	4.30 ± 0.21	1.38 ± 0.11	2.12 ± 0.17
Higher education	8.91 ± 0.55	6.50 ± 0.56	6.34 ± 0.52	5.26 ± 0.51	2.80 ± 0.31	4.19 ± 0.33	1.64 ± 0.22	1.70 ± 0.16
p-value	0.130	0.246	0.167	0.800	0.407	0.919	0.165	0.858

Table 3. *Cont.*

Variables	Starch	Fruits	Vegetables	Milk and Dairy Products	Red Meat	Poultry	Fish	Legumes
Work status								
Employed	8.59 ± 0.40	5.34 ± 0.41	4.87 ± 0.33	5.66 ± 0.38	2.93 ± 0.25	5.10 ± 0.30	1.20 ± 0.14	2.14 ± 0.21
Freelance Job	8.18 ± 0.96	4.77 ± 0.89	5.50 ± 0.89	5.07 ± 0.84	2.90 ± 0.61	3.66 ± 0.55	1.70 ± 0.37	1.85 ± 0.49
Retired	7.80 ± 0.35	6.80 ± 0.38	6.03 ± 0.35	5.59 ± 0.33	2.59 ± 0.18	3.96 ± 0.21	1.61 ± 0.13	2.04 ± 0.15
Unemployed	7.43 ± 0.65	5.13 ± 0.78	6.23 ± 0.70	5.41 ± 0.79	1.75 ± 0.33	3.66 ± 0.54	0.74 ± 0.08	1.68 ± 0.40
p-value	0.547	0.005	0.127	0.830	0.043	0.005	0.001	0.147
Income (SR)								
<2000	7.20 ± 0.77	4.71 ± 0.87	5.79 ± 0.76	5.82 ± 0.94	2.10 ± 0.45	3.56 ± 0.60	0.82 ± 0.13	1.70 ± 0.41
2000–5000	7.01 ± 0.85	3.93 ± 0.86	5.54 ± 0.77	5.55 ± 0.87	2.14 ± 0.49	4.02 ± 0.60	1.52 ± 0.42	1.63 ± 0.33
5000–7000	10.2 ± 0.97	5.65 ± 1.24	4.30 ± 0.82	7.02 ± 1.02	1.80 ± 0.27	3.87 ± 0.66	1.10 ± 0.23	2.76 ± 0.62
7000–10,000	7.42 ± 0.68	6.31 ± 0.79	5.13 ± 0.71	5.81 ± 0.71	2.11 ± 0.39	3.66 ± 0.43	0.89 ± 0.14	1.44 ± 0.20
>10,000	8.28 ± 0.29	6.28 ± 0.30	5.73 ± 0.27	5.36 ± 0.27	2.94 ± 0.17	4.65 ± 0.20	1.52 ± 0.11	2.16 ± 0.15
p-value	0.115	0.004	0.438	0.594	0.001	0.033	0.012	0.017
BMI category [2]								
Underweight	4.75 ± 0.75	5.88 ± 3.04	5.25 ± 2.98	5.13 ± 3.04	1.37 ± 0.37	4.75 ± 3.09	0.62 ± 0.23	1.50 ± 0.28
Normal weight	7.64 ± 0.53	5.76 ± 0.58	5.22 ± 0.48	5.30 ± 0.51	2.55 ± 0.29	3.71 ± 0.30	1.35 ± 0.12	1.72 ± 0.22
Overweight	7.69 ± 0.37	5.91 ± 0.38	5.88 ± 0.35	5.29 ± 0.36	2.67 ± 0.23	4.28 ± 0.25	1.57 ± 0.17	2.15 ± 0.20
Obese	8.83 ± 0.39	6.02 ± 0.43	5.44 ± 0.36	6.00 ± 0.38	2.70 ± 0.22	4.74 ± 0.29	1.17 ± 0.09	2.07 ± 0.19
p-value	0.094	0.929	0.657	0.356	0.873	0.371	0.487	0.636

[1] Dietary intake scores were calculated based on the frequency of intake of each food group and the portion size consumed each time and were rated on a 7-point scale. A score of 0 was given when no consumption was reported, a score of 0.5 when <1 serving/week was reported, a score of 1 when 1 serving/week was reported, 2 for 2–3 servings/week, 4 for 4–6 servings/week, 7 for 7 servings/weeks or 1 serving/day, and a score of 14 when ≥14 servings/week or more or ≥2 servings/day. [2] Self-reported weight and height used to calculate the BMI. The BMI categories are underweight (<18.5 kg/m^2), normal weight (18.5–24.9 kg/m^2), overweight (25.0–29.9 kg/m^2), and obese (≥30 kg/m^2). Differences between the groups were assessed via Kruskal–Wallis test. Abbreviations: SD, standard deviation; SR, Saudi Riyals; BMI body mass index.

3.4. Physical Activity Assessment

Out of 419 participants, 167 (40%) reported that they performed physical activity, with 97 participants (58%) performing physical activity at least 5 times a week. Different types of physical activity were performed by the study participants, including walking (69%), swimming (9%), resistance exercise (6%), cardio exercise (4%), cycling (3%), running (3%), yoga (2%), and other types of physical activities (4%), such as football and dancing. For those who reported not performing physical activity (60%), the majority of them (27%) stated that they did not have any specific reason for not performing physical activity (Figure 3).

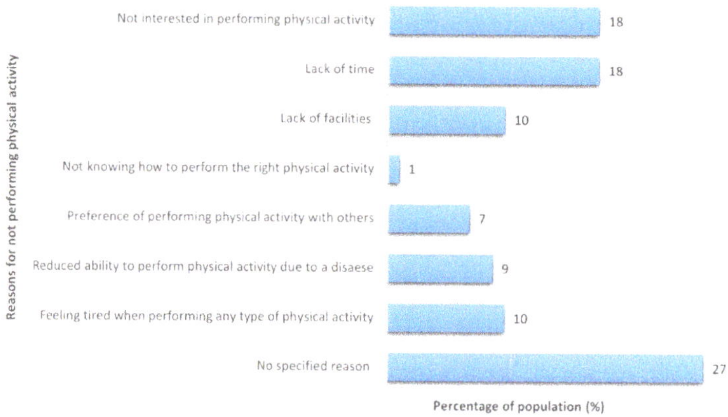

Figure 3. Reasons for not performing physical activity (n = 252).

4. Discussion

The association between diet and health status among older adults has been extensively assessed in previous studies. It has been shown that nutrition plays an important role in promoting health and preventing disease in this population [11]. However, in Saudi Arabia, the nutritional and dietary behaviors of middle-aged and older adults are not well identified. Hence, conducting this study is an important initial step to better understand the dietary patterns and behaviors of this population group, and will allow the influence of diet on the development of diseases to be better assessed in the future. Overall, the current study has shown that middle-aged and older adults living in Saudi Arabia are adopting poor dietary patterns and nutritional behaviors.

The majority of this study's participants (95%) reported that nutrition and consuming a healthy diet were important or somewhat important to them, and a large percentage (85%) rated the healthfulness of their diet as healthy or somewhat healthy. Similarly, Gille et al., who assessed eating patterns and behavior in the Swiss middle-aged and elderly population, reported that 96% of the participants said that nutrition and healthy eating were important to them, with the majority overrating their diet as healthy [12].

Most participants in this study reported that they had previously received dietary advice and reported that dietitians and physicians were the first ones to be consulted in the event of any dietary concern. Dietitians are known for their experience in nutrition and dietetics and their extensive knowledge and training in this regard, and, as such, they were expected by the public to be the first source of dietary advice [13]. Consistent with previous studies, health professionals have been shown to be the most preferred source of dietary advice [14,15]. Recent studies have reported the increased use of social media platforms for health information [16,17]. Although people usually perceive the lack of accuracy in the information obtained from the internet and social media, speed, cheapness, and ease of access could be potential factors for those looking for nutrition information [18]. This is in line with the current study's findings, which showed that a quarter of the participants accessed nutritional information by the use of social media platforms. These results suggest the importance of adopting new strategies to provide appropriate nutrition information by health professionals. For instance, virtual consultations can be used by dietitians as a more convenient and quick approach in place of face-to-face appointments.

Around half of the participants in the present study reported reading nutrition facts labels when buying new food items, with only 18% of those stating that they always read them. In contrast, a recent study was conducted among Saudi adults aged 18 years and over that showed that 62% of the studied population reported using nutrition facts labels when buying food products [19]. A possible explanation for the lower percentage of nutrition fact label usage in this study is due to the age of the recruited participants. In fact, 54% of this study participants were aged 55 years old and older, while the majority of the other study participants were younger, with only 7% of them aged 56 years or older. The influence of aging could be one of the factors that affect participants' ability to understand or to interpret the information provided in the labels, which, in turn, will reduce the use of the information of food labels [20]. The present study supports this explanation and showed that unclear fonts and designs and difficulty in understanding the content of nutrition facts labels limited the use of the nutrition facts labels by a large number of the participants. This highlights the need to use other formats that might enhance consumers' comprehension of the information provided on nutrition facts labels. Additionally, campaigns to increase awareness of how to use the information provided in the labels could be conducted, thereby guiding the consumer toward healthier and better food choices.

One of the main aims of the current study was to assess the dietary patterns of the study population based on the frequency and quantity of consumption of different food groups. This study showed that the average fruit and vegetable consumption was six servings a week each, which is much lower than the intake recommended by the World Health Organization (WHO), which recommends at least five servings of fruits and vegetables to be consumed on a daily basis [21]. The low consumption of fruits and vegetables in this study population could be due to several barriers that might affect their consumption rates as they age. These factors include social isolation, poor dental health, and increased disease susceptibility [22]. Age was one of the factors affecting the consumption of fruits and vegetables in this study, with higher intakes at older ages. Similarly, it has been previously shown that individuals tend to increase their fruit and vegetable consumption as they age [22–24]. Fruits in Saudi Arabia are costly, and this could deter a large number of the population from purchasing fruits on a regular basis, particularly those with low monthly incomes [9,25]. This was reported in this study, where those with an income of higher than 10,000 riyals per month (around $2600), consumed significantly more servings of fruits compared with those with a lower income. An American study that assessed the effect of income on older adults' eating patterns showed that those with low and medium incomes consumed significantly fewer fruits than those with higher incomes [26]. For that reason, it was previously estimated that increased income along with reduced fruit prices would likely lead to an increase in the rate of fruit consumption in the world's elderly population [27].

Recent studies have highlighted the importance of protein adequacy for preserving muscle mass and function [28]. However, the adverse impact of meat overconsumption, red meat in particular, has been reported and it has been shown that moderation is the key for healthy aging [29]. Participants in this study exceeded the WHO-recommended weekly intake of red meat and poultry [21] while fish intake was below the WHO's recommendations with an average weekly intake of three servings of red meat, four servings of poultry, and one serving of fish. Gender significantly influenced red meat, poultry, and fish consumption in the current study, with higher scores for men. This finding is in line with Saudi studies that have assessed dietary pattern in Saudi adults and shown that red meat and fish consumption was higher in men compared to women [9,30]. This could be explained by the fact that women often show greater dietary knowledge and therefore make better food choices than men [31,32]. Moreover, intakes of red meat, poultry, and fish were influenced by other factors in this study population, including monthly income. The average consumption was higher in those with higher monthly incomes, and this was similarly reported previously [33,34].

In Saudi Arabia, the prevalence of inactivity has always been a challenge to public health, as was shown in a study that assessed the performance of physical activity among adults aged 15 years old and over living in Saudi Arabia and showed that performance of physical activity was low and decreased with age, with around 88% of the population aged 45 and over reporting being inactive [35]. The present study supports this finding, showing a low level of physical activity, with 60% of the study participants stating that they did not perform physical activity.

Despite the importance of the present study in providing insight into nutritional behaviors and dietary patterns in middle-aged and older adults in Saudi Arabia, the study has some limitations. To allow for larger sample size recruitment and to obtain responses from various demographic populations, data in the current study were collected by distributing the questionnaire electronically. However, this approach could introduce a slight bias in the study results as it might not represent the actual sociodemographic structure of the entire population in Saudi Arabia. Another limitation is that participants in this study were asked to self-report the frequencies and quantities of different food groups' intake rather than measuring the actual dietary intake, and this could increase the chance of intake misreporting.

5. Conclusions

Adopting healthy habits has been linked with improved health and wellbeing as people get older. Middle-aged and older adults living in Saudi Arabia seem to have poor dietary patterns and nutritional behaviors. While the majority of this study population considered nutrition to be important to them, a great percentage of them reported not using nutrition facts labels, having poor consumption levels of fruits, vegetables, fish, and legumes, and engaging in little physical activity. Therefore, implementing effective strategies to enhance knowledge of the importance of nutrition among the older population is needed. Nutrition education and guidance are warranted for middle-aged and older adults to help them improve their dietary choices and overall lifestyles as they get older. Likewise, there is a need for further studies assessing the actual dietary intake of different food group populations in Saudi Arabia with larger sample sizes, as well as comparisons with Saudi dietary recommendations.

Funding: This research received no external funding.

Institutional Review Board Statement: The study was conducted according to the guidelines of the Declaration of Helsinki and approved by the Unit of the Biomedical Ethics Research Committee at King Abdulaziz University (Jeddah, Saudi Arabia) (reference No. 426-21).

Informed Consent Statement: Informed consent was obtained from all subjects involved in the study.

Data Availability Statement: The data used to support the findings of this study are available from the corresponding author upon request.

Acknowledgments: The authors would like to thank Areej Alkhaldy, Manal Naseeb, and Manar Bakhsh from the Clinical Nutrition Department at King Abdulaziz University for reviewing the study questionnaire.

Conflicts of Interest: The author declares no conflict of interest.

References

1. World Health Organization. Ageing and Health. Available online: https://www.who.int/news-room/fact-sheets/detail/ageing-and-health (accessed on 11 November 2021).
2. United Nations. UNData: A World of Information. Available online: https://data.un.org/en/iso/sa.html (accessed on 21 August 2021).
3. Colloca, G.; Santoro, M.; Gambassi, G. Age-Related Physiologic Changes and Perioperative Management of Elderly Patients. *Surg. Oncol.* **2010**, *19*, 124–130. [CrossRef] [PubMed]

4. Frontera, W.R. Physiologic Changes of the Musculoskeletal System with Aging: A Brief Review. *Phys. Med. Rehabil. Clin. N. Am.* **2017**, *28*, 705–711. [CrossRef]
5. Franco, O.H.; Karnik, K.; Osborne, G.; Ordovas, J.M.; Catt, M.; Van Der Ouderaa, F. Changing Course in Ageing Research: The Healthy Ageing Phenotype. *Maturitas* **2009**, *63*, 13–19. [CrossRef]
6. Black, M.; Bowman, M. Nutrition and Healthy Aging. *Clin. Geriatr. Med.* **2020**, *36*, 655–669. [CrossRef]
7. Roberts, S.B.; Silver, R.E.; Das, S.K.; Fielding, R.A.; Gilhooly, C.H.; Jacques, P.F.; Kelly, J.M.; Mason, J.B.; McKeown, N.M.; Reardon, M.A.; et al. Healthy Aging—Nutrition Matters: Start Early and Screen Often. *Adv. Nutr.* **2021**, *12*, 1438–1448. [CrossRef]
8. Chiuve, S.; Fung, T.; Rimm, E.; Hu, F.; McCullough, M.; Wang, M.; Stampfer, M.; Willett, W. Alternative Dietary Indices Both Strongly Predict Risk of Chronic Disease. *J. Nutr.* **2012**, *142*, 1009–1018. [CrossRef] [PubMed]
9. Moradi-Lakeh, M.; El Bcheraoui, C.; Afshin, A.; Daoud, F.; Almazroa, M.A.; Al Saeedi, M.; Basulaiman, M.; Memish, Z.A.; Al Rabeeah, A.A.; Mokdad, A.H. Diet in Saudi Arabia: Findings from a Nationally Representative Survey. *Public Health Nutr.* **2017**, *20*, 1075–1081. [CrossRef]
10. General Authority for Statistics Kingdom of Saudi Arabia. Population by Age Groups, and Gender Mid Year 2020. Available online: https://www.stats.gov.sa/en/6768 (accessed on 12 November 2021).
11. Kiefte-De Jong, J.C.; Mathers, J.C.; Franco, O.H. Nutrition and Healthy Ageing: The Key Ingredients. *Proc. Nutr. Soc.* **2014**, *73*, 249–259. [CrossRef] [PubMed]
12. Gille, D.; Bütikofer, U.; Chollet, M.; Schmid, A.; Altintzoglou, T.; Honkanen, P.; Stoffers, H.; Walther, B.; Piccinali, P. Nutrition Behavior of the Middle-Aged and Elderly: Compliance with Dietary Recommendations of the Food Pyramid. *Clin. Nutr.* **2016**, *35*, 638–644. [CrossRef]
13. Andersen, D.; Baird, S.; Bates, T.; Chapel, D.L.; Cline, A.D.; Ganesh, S.N.; Garner, M.; Grant, B.L.; Hamilton, K.K.; Jablonski, K.; et al. Academy of Nutrition and Dietetics: Revised 2017 Standards of Practice in Nutrition Care and Standards of Professional Performance for Registered Dietitian Nutritionists. *J. Acad. Nutr. Diet.* **2018**, *118*, 132–140.e15. [CrossRef]
14. Winter, J.E.; Mcnaughton, S.A.; Nowson, C.A. Older Adults' Attitudes to Food and Nutrition: A Qualitative Study. *J. aging Res. Clin. Pract.* **2016**, *5*, 114–119. [CrossRef]
15. Ball, L.; Desbrow, B.; Leveritt, M. An Exploration of Individuals' Preferences for Nutrition Care'from Australian Primary Care Health Professionals. *Artic. Aust. J. Prim. Health* **2013**, *20*, 113–120. [CrossRef]
16. Dumas, A.A.; Lapointe, A.; Desroches, S. Users, Uses, and Effects of Social Media in Dietetic Practice: Scoping Review of the Quantitative and Qualitative Evidence. *J. Med. Internet Res.* **2018**, *20*. [CrossRef]
17. Adamski, M.; Truby, H.; Klassen, K.M.; Cowan, S.; Gibson, S. Using the Internet: Nutrition Information-Seeking Behaviours of Lay People Enrolled in a Massive Online Nutrition Course. *Nutrients* **2020**, *12*, 750. [CrossRef]
18. Cash, T.; Desbrow, B.; Leveritt, M.; Ball, L. Utilization and Preference of Nutrition Information Sources in Australia. *Heal. Expect.* **2015**, *18*, 2288–2295. [CrossRef]
19. Arfaoui, L.; Alkhaldy, A.; Alareeshi, A.; Alsaadi, G.; Alhendi, S.; Alghanmi, A.; Alghafari, W.; Assidi, M. Assessment of Knowledge and Self-Reported Use of Nutrition Facts Labels, Nutrient Content, and Health Claims among Saudi Adult Consumers. *J. Multidiscip. Healthc.* **2021**, *14*, 2959. [CrossRef]
20. Miller, L.M.S.; Applegate, E.; Beckett, L.A.; Wilson, M.D.; Gibson, T.N. Age Differences in the Use of Serving Size Information on Food Labels: Numeracy or Attention? *Public Health Nutr.* **2017**, *20*, 786. [CrossRef] [PubMed]
21. World Health Organization. Healthy Diet. Available online: https://apps.who.int/iris/bitstream/handle/10665/325828/EMRO PUB_2019_en_23536.pdf (accessed on 10 July 2022).
22. Sahyoun, N.R.; Zhang, X.L.; Serdula, M.K. Barriers to the Consumption of Fruits and Vegetables among Older Adults. *J. Nutr. Elder.* **2006**, *24*, 5–21. [CrossRef]
23. Nicklett, E.J.; Kadell, A.R. Fruit and Vegetable Intake among Older Adults: A Scoping Review. *Maturitas* **2013**, *75*, 305–312. [CrossRef]
24. Drewnowski, A.; Shultz, J.M. Impact of Aging on Eating Behaviors, Food Choices, Nutrition, and Health Status. *J. Nutr. Health Aging* **2001**, *5*, 75–79.
25. El Bcheraoui, C.; Basulaiman, M.; AlMazroa, M.A.; Tuffaha, M.; Daoud, F.; Wilson, S.; Al Saeedi, M.Y.; Alanazi, F.M.; Ibrahim, M.E.; Ahmed, E.M.; et al. Fruit and Vegetable Consumption among Adults in Saudi Arabia, 2013. *Nutr. Diet. Suppl.* **2015**, *7*, 41–49. [CrossRef]
26. Bowman, S. Low Economic Status Is Associated with Suboptimal Intakes of Nutritious Foods by Adults in the National Health and Nutrition Examination Survey 1999-2002. *Nutr. Res.* **2007**, *27*, 515–523. [CrossRef]
27. Muhammad, A.; D'Souza, A.; Meade, B.; Micha, R.; Mozaffarian, D. How Income and Food Prices Influence Global Dietary Intakes by Age and Sex: Evidence from 164 Countries. *BMJ Glob. Health* **2017**, *2*, e000184. [CrossRef]
28. Carbone, J.W.; Pasiakos, S.M. Dietary Protein and Muscle Mass: Translating Science to Application and Health Benefit. *Nutrients* **2019**, *11*, 1136. [CrossRef]
29. Kouvari, M.; Tyrovolas, S.; Panagiotakos, D.B. Red Meat Consumption and Healthy Ageing: A Review. *Maturitas* **2016**, *84*, 17–24. [CrossRef]
30. Alsufiani, H.M.; Kumosani, T.A.; Ford, D.; Mathers, J.C. Dietary Patterns, Nutrient Intakes, and Nutritional and Physical Activity Status of Saudi Older Adults: A Narrative Review. *J. Aging Res. Lifestyle* **2015**, *4*, 2–11. [CrossRef]

31. Wardle, J.; Haase, A.M.; Steptoe, A.; Nillapun, M.; Jonwutiwes, K.; Bellisle, F. Gender Differences in Food Choice: The Contribution of Health Beliefs and Dieting. *Ann. Behav. Med.* **2004**, *27*, 107–116. [CrossRef]
32. DicksonSpillmann, M.; Siegrist, M. Consumers' Knowledge of Healthy Diets and Its Correlation with Dietary Behaviour. *J. Hum. Nutr. Diet.* **2011**, *24*, 54–60. [CrossRef]
33. Trondsen, T.; Braaten, T.; Lund, E.; Eggen, A.E. Consumption of Seafood—the Influence of Overweight and Health Beliefs. *Food Qual. Prefer.* **2004**, *15*, 361–374. [CrossRef]
34. Burger, J.; Gochfeld, M.; Batang, Z.; Alikunhi, N.; Al-Jahdali, R.; Al-Jebreen, D.; Aziz, M.A.M.; Al-Suwailem, A. Fish Consumption Behavior and Rates in Native and Non-Native People in Saudi Arabia. *Environ. Res.* **2014**, *133*, 141–148. [CrossRef]
35. Alqahtani, B.A.; Alenazi, A.M.; Alhowimel, A.S.; Elnaggar, R.K. The Descriptive Pattern of Physical Activity in Saudi Arabia: Analysis of National Survey Data. *Int. Health* **2021**, *13*, 232–239. [CrossRef]

nutrients MDPI

Article

Health-Promoting Behaviors among Older Adults with Noncommunicable Diseases in Rural and Urban Areas during the New Normal Post-COVID-19 Era: A Structural Equation Modeling Analysis

Wanich Suksatan [1], Supat Teravecharoenchai [1,*] and Jintana Sarayuthpitak [2]

[1] Department of Public Health, Faculty of Liberal Arts, Krirk University, Bangkok 10220, Thailand
[2] Department of Curriculum and Instruction, Faculty of Education, Chulalongkorn University, Bangkok 10330, Thailand
* Correspondence: supatte.kru@gmail.com

Abstract: This study aimed to develop and test a causal relationship among perceived self-efficacy (PSE), health literacy (HL), access to COVID-19 preventive material (ACPM), social networks (SN), and health-promoting behaviors (HPBs). Multistage stratified random sampling was used to recruit 250 older adults with noncommunicable diseases (NCDs) from Thai urban and rural communities. The data were collected with self-reported questionnaires. Data analyses used descriptive statistics and structural equation modeling. The results indicated that participants in urban communities had higher PSE, ACPM, HL, SN, and HPBs than rural participants. The fitness parameters of the modified model (χ^2 = 71.936, df = 58, p-value = 0.103, χ^2/df = 1.240; root mean square error of approximation (RMSEA) = 0.031; standardized root mean square residual (SRMR) = 0.042; goodness of fit index (GFI) = 0.964; normed-fit index (NFI) = 0.964; comparative fit index (CFI) = 0.993) indicated its suitability as the research model. HPBs were directly positively influenced by PSE (β = 0.40, $p < 0.001$), ACPM (β = 0.24, $p < 0.001$), HL (β = 0.19, $p < 0.01$), and SN (β = 0.01, $p < 0.05$). Therefore, taking all predicting variables together could explain 81.0% of the variance in HPBs. Multidisciplinary healthcare teams could use these findings to establish proper interventions or healthcare activities to increase HPBs among older adults, particularly in this era of the "new normal".

Keywords: COVID-19 pandemic; health-promoting behaviors; older adults; nutrition; NCDs

1. Introduction

The coronavirus disease 2019 (COVID-19) began a pandemic at the end of December 2019 and has caused severe damage to cities around the world [1]. Vaccination with the accelerated COVID-19 vaccine has been mandated worldwide to prevent the spread of COVID-19 [2,3]. However, the development and spread of the Delta and Omicron variants of COVID-19 could reduce the effectiveness of the vaccine and jeopardize efforts to contain the pandemic [4,5]. After approximately 3 years, 642 million people have had confirmed cases of COVID-19, and there have been 6.62 million deaths across the globe. In Thailand, there has been a cumulative total of 4.7 million people with confirmed cases and 33,285 deaths due to COVID-19 [6]. Since it began, the COVID-19 pandemic has had an impact on all people and in diverse areas, including the economy, society, and health-related well-being [7].

Numerous studies have found that older adults were disproportionately affected by COVID-19 during the global Omicron wave [8]. Older adult populations were regarded as being at greater risk of developing serious illness from COVID-19 than were younger people [9]. Prior risk analyses also revealed that older adults with noncommunicable diseases (NCDs), such as chronic kidney disease, diabetes, cancer, and heart conditions, might be more susceptible to developing severe illness [9]. However, several countries, including Thailand, have attempted to prevent and control the spread of COVID-19 by promoting healthy lifestyle behaviors, such as hand washing, mask wearing, social distancing, and avoiding crowded places [10]. In the "new normal" of the post-COVID-19 period, the Thai government has helped people realize that healthy lifestyles are the key to resilience in the fight against COVID-19 and numerous other health threats [10]. Health promotion aims to improve the health of all generations at all times, rather than just focusing on emergencies; however, COVID-19 may aid in driving the message home [11].

Health-promoting behaviors (HPBs) are particularly important for older adults with NCDs because they are effective methods of maintaining appropriate self-care. They also actively promote health-related well-being and better quality of life [12]. Pender's health promotion model is regarded as an essential tool for determining health status, preventing disease, and enhancing health and well-being throughout a person's lifetime [13], especially for epidemic control measures (e.g., social distancing, self-isolating, or home-sheltering) that disrupt the delivery of various types of health services. A global survey by the World Health Organization revealed that health treatments and prevention services for people with NCDs were significantly reduced or discontinued due to the COVID-19 pandemic [14]. A sustained and prolonged epidemic may affect HPBs, thereby diminishing the efficacy of COVID-19 controls and making it more difficult to maintain the lifestyles and well-being of older adults with NCDs [15,16]. These factors related to recent events have led to older adults becoming more aware of individual and public health. Improving HPBs by maintaining good health through stress management and a diet of foods that boost immunity can help protect this population from contracting this and other viruses.

Although several studies assessed HPBs during the first and second waves of the COVID-19 outbreak, most previous studies focused on the HPBs of hospitalized patients [17] and community-dwelling older adults [12,18]. Empirically, prior studies reported that factors associated with HPBs among older adults include perceived self-efficacy (PSE) [12,19], health literacy (HL) [20,21], access to COVID-19 preventive material (ACPM) [20], and social networks (SN) [18]. Consequently, these elements need to be accounted for when considering HPBs among this population. To the best of our knowledge, no study has captured any causal relationships or conducted a comparative study of urban and rural communities of older adults with NCDs with regard to HPBs in the new normal post-COVID-19 period. Consequently, this gap must be filled through the discernment of a causal relationship of HPBs that addresses these specific variables in older adults with NCDs. We selected the factors that can be modified by healthcare providers, including PSE, HL, ACPM, and SN, as the influencing factors of HPBs. With those selections, this study aimed to compare these variables between rural and urban older adults with NCDs as well as develop and test a hypothesized causal model of HPBs among older adults with NCDs who lived in rural and urban Thai communities post-COVID-19. In this new normal, it is necessary to have a better understanding of these factors and how they affect HPBs and other people, especially when there are differences, in order to guide and establish interventions that improve health and well-being among older adults in different residential areas.

2. Materials and Methods

2.1. Research Design

The cross-sectional research design allowed this study to determine the consistency of the causal relationship model for PSE, HL, ACPM, SN, and HPBs using empirical data and examine the effects of these factors on HPBs among older adults with NCDs (see Figure 1). The data from "Determinants of the Health-Promoting Behaviors among Community-Dwelling Older Adults with Non-Communicable Diseases during the New Normal Post-COVID-19 Era" was used to obtain the aims of the present study.

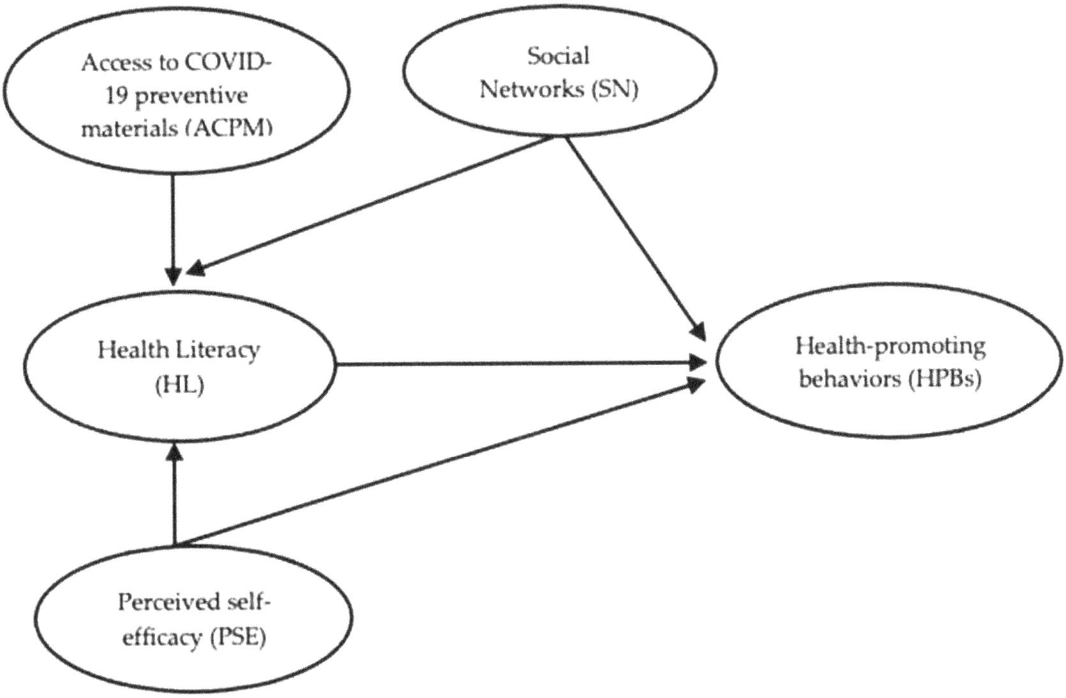

Figure 1. Research framework for the study.

2.2. Setting and Participants

Thailand defines an older adult as a person aged 60 years or older [22], and the urban and rural communities are classified according to residential areas [23]. The participants in this study were older adults with NCDs from Ubon Ratchathani city municipality (urban communities) and Huaruea and Nong Khon sub-districts (rural communities) in Ubon Ratchathani province in northeast Thailand and were recruited for this study via multistage stratified and simple random sampling selections. The study sample comprised (1) Thai older adults aged 60 years or more who had been (2) diagnosed with one or more NCD (i.e., vascular disease, heart disease, chronic obstructive pulmonary disease, diabetes, hypertension, cancer, and obesity) and (3) who could understand and communicate in Thai. Participants who were unable to complete the questionnaire due to visual/sensory/auditory abnormalities or who were unwilling to participate in this study were excluded. The number of possible sample sizes needed for a study with the structural equation model ranges from 100 to 800 [24]. We determined a suitable sample size using G*Power software version 3.1.9 [25] and determined that the effect size was 0.15 [26], with an alpha level of 0.05 and a power of 0.80. We added 30% to counterbalance any incomplete questionnaires and missing data, resulting in a final sample size of 250.

2.3. Research Instruments

The following six research instruments were used to collect the data:

Sociodemographic Data Form. This form, developed by the researchers, included four multiple choice or open-ended questions asking about the participant's age, sex, monthly income, and type of NCD.

Health Literacy Scale (HLS). This measure was originally developed in Thai by the Health Education Division [27]. It comprises 10 items, and each item is answered on a

4-point ratings scale, with scores ranging from 1 (strongly disagree) to 5 (strongly agree). The total score is 10–50, and mean scores are divided into four levels: bad (10–19), fair (20–29), good (30–39), and very good (40–50), with higher scores indicating higher levels of literacy on health [27]. The psychometric properties of the HLS were tested and found to be valid and reliable [27]; the Cronbach's alpha coefficient of this scale was 0.90 for this study.

Lubben Social Network Scale (LSNS-6). This measure was originally developed by Lubben et al. [28] and comprises six items with two subscales: family and friendships. Each item is answered on a 6-point rating scale with scores ranging from 0 (none) to 5 (9 or more). The total score is 0–30, and mean scores are divided into two levels: social isolation (0–12) and social engagement (13–30), with higher scores indicating more social engagement [28]. After the construct validity of the LSNS-6 was tested using confirmatory factor analysis (CFA), six items with two subscales remained and fit with the empirical data [28]. The Cronbach's alpha coefficient of the LSNS-6 in this study was 0.88.

Self-Rated Abilities Scale for Health Practice (SRAHP). This measure was originally developed by Becker et al. [29]. It comprises 28 items with 4 subscales: nutrition, stress management, exercise, and health practice. Each item is answered on a 5-point ratings scale, with scores ranging from 0 (not at all) to 4 (completely). The total score is 0–112, and mean scores are divided into three levels: low (0–37), fair (38–74), and high (75–112), with higher scores indicating a higher level of self-efficacy [29]. After the construct validity of SRAHP was tested using CFA, four factors remained and fit with the empirical data [29]. The Cronbach's alpha coefficient of SRAHP in this study was 0.95.

Access to COVID-19 Preventive Materials (ACPMS). This measure was originally developed in Thai by Yodmai et al. [20]. It comprises 5 items, with each item answered on a 3-point rating scale. Scores range from 0 (no or not sure) to 1 (yes), with a total score range of 0–5. The mean scores are divided into two levels: bad (0–2) and good (3–5), with higher scores indicating greater access to COVID-19 preventive materials [20]. Psychometric properties of ACPMS were tested and found valid and reliable [20]. The Cronbach's alpha coefficient of ACPMS in this study was 0.76.

Health-Promoting Behaviors Scale (HPBS). This measure was originally developed by the Thai Health Education Division [27]. It comprises 19 items with 7 subscales: nutrition, exercise, smoking, alcohol drinking, stress management, rational drug use, and preventing COVID-19 infection. Each item is answered on a 5-point rating scale, with scores ranging from with 1 (not at all) to 5 (completely). The scores of HPBs are divided into four levels: bad (<60%), fair (60–69%), good (70–80%), and very good (>80%), with higher scores represent more HPBs. The Cronbach's alpha coefficient of HPBS in this study was 0.75.

2.4. Data Collection

After submitting the required permission to participate and informed consent, eligible participants were able to access the self-reported questionnaire between 10 September and 10 November 2022 during the "new normal" post-COVID-19 era. Participants took approximately 30–40 min to answer all of the questions, and a total of 250 participants completed the questionnaire (100%).

2.5. Data Analyses

Statistical Package for Social Sciences (SPSS) version 25.0 and AMOS (analysis of moment structure) software were used for data analyses. The demographic data and all variables were assessed using descriptive statistics. Pearson's correlation and Spearman's rank correlation coefficient were used to examine the relationships between the measured variables. The independent t-test was used to compare the differences between rural and urban older adults with NCDs with regard to PSE, HL, ACPM, SN, and HPBs. Additionally, a structural equation model (SEM) was used to determine the causal relationships. Prior to the data analysis, all relevant assumptions were met. The fit of the hypothesized model was assessed based on several criteria, including (a) chi-square test (χ^2, $p > 0.05$); (b) normed chi-square (χ^2/df) with the desired value of <3; (c) the value of the root mean square error of approximation (RMSEA), which was ≤0.05; (d) the value of the comparative fit index (CFI), which was ≥0.90; (e) the normed-fit index (NFI), which was ≥ 0.90; (f) the value of standardized root mean square residual (SRMR), which was ≤0.08; and (g) the goodness-of-fit index (GFI), which was ≥0.90 [30]. The significance level was set at $p < 0.05$ for all analyses.

3. Results

3.1. Characteristics of Participants

The demographic data for the 250 participants revealed the following: 63.60% were female, 52.80% were aged 60–69 years (mean = 69.61, SD ± 7.47), 82.80% were generally uniformly spread on monthly income of less than USD 143 (mean = USD 132.14, SD ± 239.20), and 65.60% participants were living with hypertension. In addition, participants' ages, education levels, and monthly income were positively significant with HPBs ($p < 0.05$), as shown in Table 1.

Table 1. Participants' demographic data.

Characteristic	Health-Promoting Behaviors			*p*-Value
	Urban (*n* = 125)	Rural (*n* = 125)	Total (*n* = 250)	
	n (%)	*n* (%)	*n* (%)	
Sex				
Female	92 (73.60)	67 (53.60)	159 (63.60)	0.446
Age (years): Mean ± SD	70.59 ± 7.44	68.63 ± 7.41	69.61 ± 7.47	0.018
60–69	55 (44.00)	77 (61.60)	132 (52.80)	
70–79	54 (43.20)	34 (27.20)	88 (35.20)	
>80	16 (12.80)	14 (11.20)	30 (12.00)	
Monthly income (US dollars): Mean ± SD	193.58 ± 321.05	70.69 ± 64.91	132.14 ± 239.20	0.026
<143	92 (73.60)	115 (92.00)	207 (82.80)	
144–286	10 (8.00)	10 (8.00)	20 (8.00)	
286–429	6 (4.80)	0 (0.00)	6 (2.40)	
>430	17 (13.60)	0 (0.00)	17 (6.80)	
Type of NCDs (Yes) *				0.176
Heart disease	9 (7.20)	11 (8.80)	20 (8.00)	
Vascular disease	15 (12.00)	8 (6.40)	23 (9.20)	
Diabetes	67 (53.60)	46 (36.80)	113 (45.20)	
Hypertension	82 (65.60)	82 (65.60)	164 (65.60)	
Cancer	1 (0.80)	1 (0.80)	2 (0.80)	
Chronic obstructive pulmonary disease	0 (0.00)	3 (2.40)	3 (1.20)	
Obesity	9 (7.20)	2 (16.0)	11 (4.40)	

NCDs = noncommunicable diseases, SD = standard deviation. * This measure permitted multiple answers.

3.2. Perceived Self-Efficacy, Health Literacy, Access to COVID-19 Preventive Material, Social Networks, and Health-Promoting Behaviors among Urban and Rural Older Adults with NCDs

In this study, the participants had high levels of PSE (mean = 76.48, SD ± 17.55), and the SN indicated higher social engagement levels (mean = 15.08, SD ± 5.59). HL (mean = 37.28, SD ± 6.37), ACPM (mean = 37.28, SD ± 6.37), and HPBs (mean = 67.44, SD ± 7.51) were also all at good levels. In comparisons of the mean of PSE, HL, ACPM, SN, and HPBs, it was found that PSE, HL, ACPM, SN, and HPBs were significantly higher among participants who lived in urban communities than those in rural ones. We also found that all domains of PSE and HPBs were significantly higher for urban dwellers than rural respondents (see Table 2).

Table 2. Perceived self-efficacy, health literacy, access to COVID-19 preventive material, social networks, and health-promoting behaviors among urban and rural older adults with NCDs.

Variables	Interpretation	Urban (n = 125)		Rural (n = 125)		Total (n = 250)		p-Value
		Mean	*SD*	Mean	*SD*	Mean	*SD*	
Perceived self-efficacy (PSE)	High	80.54	17.7	72.42	16.50	76.48	17.55	<0.001
Nutrition self-efficacy	High	21.09	4.92	19.67	4.17	20.38	4.61	0.014
Stress management self-efficacy	High	20.12	4.24	17.93	4.28	19.02	4.39	<0.001
Exercise self-efficacy	Fair	18.55	5.63	15.12	5.87	16.83	5.99	<0.001
Health practice self-efficacy	High	20.77	4.47	19.69	4.70	20.23	4.61	0.064
Health literacy (HL)	Good	37.50	6.78	37.06	5.94	37.28	6.37	0.586
Access to COVID-19 preventive material (ACPM)	Good	4.46	0.92	3.08	1.30	4.14	1.18	<0.001
Social networks	More social engagement	15.16	5.93	14.99	5.24	15.08	5.59	0.813
Health-promoting behaviors (HPBs)	Good	68.13	7.16	66.73	7.81	67.44	7.51	0.141
Nutrition	Good	18.91	2.79	18.28	2.97	18.59	2.89	0.084
Exercise	Fair	6.11	2.12	6.08	1.84	6.09	1.98	0.899
Smoking	Very Good	9.19	1.77	8.69	1.94	8.94	1.87	0.036
Alcohol drinking	Very Good	4.61	0.90	4.53	0.98	4.57	0.94	0.505
Stress management	Fair	5.23	1.04	5.78	1.28	5.50	1.19	<0.001
Rational drug use	Very Good	9.89	1.85	9.67	2.21	9.78	2.04	0.387
Preventive COVID-19 infection	Very Good	14.17	1.67	13.68	1.57	13.93	1.64	0.019

3.3. Structural Model

An SEM was used to evaluate the causal relationships based on the constructed framework and the null hypothesis. The results revealed that the obtained fit indices were χ^2 = 556.163, df = 85, p-value = 0.000, χ^2/df = 6.543; RMSEA = 0.149; SRMR = 0.166; GFI = 0.773; NFI = 0.719; CF = 0.749. We found the causal relationship model indicated that some statistical criteria were at unacceptable levels based on the empirical data [30]. Therefore, the model fit needed to be modified by adjusting the errors of several observed variables to allow relationships between them to increase the fit index values to an acceptable level.

After the adjustment, the model fit of HPB indices were acceptable, with the empirical data of χ^2 = 71.936, df = 58, p-value = 0.103, χ^2/df = 1.240; RMSEA = 0.031; SRMR = 0.042; GFI = 0.964; NFI = 0.964; and CFI = 0.993. The model explained 81.0% of the total variance in HPBs among the older adults with NCDs in our study (see Table 3 and Figure 2).

Table 3. The direct, indirect, and total effects among the variables in the study.

Dependent Variables	R^2	Effects	Independent Variables: ACPM	PSE	SN	HL
HL	0.72	DE	−0.03 * (−0.85)	0.81 *** (13.41)	0.11 * (2.14)	–
		IE	–	–	–	–
		TE	−0.03 * (−0.85)	0.81 *** (13.41)	0.11 * (2.14)	–
HPBs	0.81	DE	0.24 *** (3.55)	0.40 *** (4.32)	0.01 * (0.91)	0.19 ** (2.36)
		IE	−0.01 * (−0.09)	0.15 ** (1.61)	0.02 * (0.39)	–
		TE	0.23 *** (3.46)	0.55 *** (5.92)	0.03 * (0.51)	0.19 ** (2.36)
χ^2 = 71.936, df = 58, p-value = 0.103, χ^2/df = 1.240; RMSEA = 0.031; SRMR = 0.042; GFI = 0.964; NFI = 0.964; CFI = 0.993						

ACPM = access to COVID-19 preventive materials, PSE = perceived self-efficacy, SN = social networks, HL = health literacy, HPBs = health-promoting behaviors, DE = direct effect, IE = indirect effect, TE = total effect. * $p < 0.05$, ** $p < 0.01$, *** $p < 0.001$.

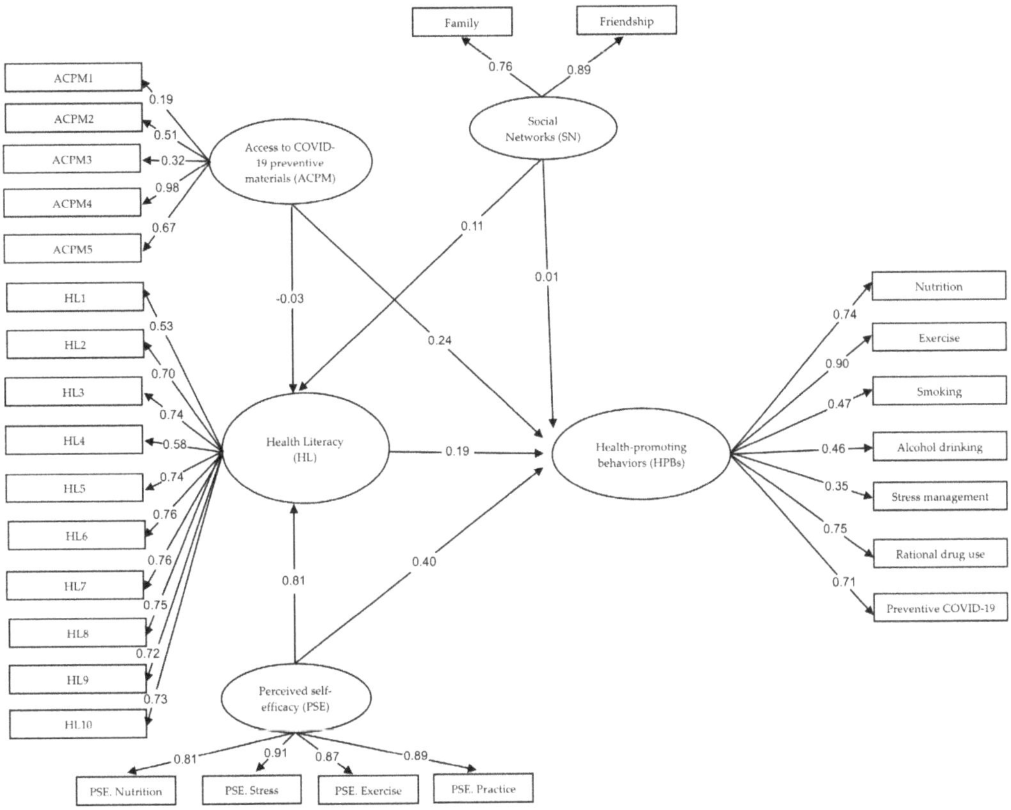

Figure 2. Causal relationship model of HPBs among the older adults with NCDs in our study.

The standardized coefficients of the final model of HPBs showed that HPBs among older adults with NCDs was directly influenced positively by PSE ($\beta = 0.40$, $p < 0.001$), ACPM ($\beta = 0.24$, $p < 0.001$), HL ($\beta = 0.19$, $p < 0.01$), and SN ($\beta = 0.01$, $p < 0.05$). Therefore, all of the predicting variables were kept in the model and, all together, could explain 81.0% of the variance in HPBs.

Regarding indirect causal effects on HPBs, the results of the final model revealed that the three predictors of PSE ($\beta = 0.15$, $p < 0.01$), SN ($\beta = 0.02$, $p < 0.05$), and ACPM ($\beta = -0.01$, $p < 0.05$) indirectly affected HPBs via HL. The finding showed that HL was a mediator of HPBs among older adults with NCDs. In addition, the three variables altogether explained 72.0% of the variance in HL.

4. Discussion

To the best of our knowledge, this is the first study to directly examine the associations between the levels of PSE, HL, ACPM, SN, and HPBs and compare those variables in older adults with NCDs. This study also aimed to develop and test a hypothesized causal model of HPBs among older adults with NCDs who lived in Thai rural and urban communities during the "new normal" post-COVID-19 era.

In this study, we also determined the causal relationship, and our findings revealed that PSE is the most influential factor affecting HPBs among older adults with NCDs. This is because PSE increases an individual's self-confidence and self-care skills; thus, people with higher PSE can improve their HPBs [13]. The results of this study are consistent with prior studies reporting PSE influences on HPBs among older adults in Thailand [12], Korea [21], Indonesia [31], and the United States [32]. PSE is a significant factor in improving motivation and, thereby, engagement in HPBs among older adults with NCDs. This means that individuals with higher PSE scores are better able to motivate themselves to engage regularly in HPBs [33]. A recent systematic review of self-efficacy and self-care among people with hypertension revealed that there were 21 studies across the globe, including in Africa, Asia, Australia, Europe, the Middle East, and the United States, reporting that higher PSE was related to self-care behaviors, such as physical activity, dietary changes, and medication adherence [34]. The findings also revealed significantly higher PSE scores in urban communities than in rural areas. The current findings are in line with previous studies that older adults who have a high level of PSE could improve their ability to perform HPBs. In fact, it has been found that older Thai adults with hypertension who live in urban areas have higher PSE and HPB scores than their peers in rural areas [12].

Additionally, we found that HL was directly affected by HPBs among older adults with NCDs. These findings support Do and colleagues' multi-institutional study that examined HL and health behaviors among older adults during the COVID-19 pandemic [35]. They found that HL was associated with health-related behaviors, which included 8% more healthy eating behaviors (95% CI: 1.04–1.13), 4% more physical activity (95% CI: 1.01–1.08), and 9% less depression (95% CI: 0.87–0.94) [35]. From Nutbeam's perspective, an individual's HL is defined as their ability to comprehend, access, and select health information based on their attitudes and motivations for appropriate self-care and HPBs [36,37]. A prior study supports our findings showing that HL was associated with HPBs and it was the combination of all dimensions of HL—comprehension, accessibility, reading skills, evaluation, and decision-making—and their behaviors could explain 58.0% of the variance in HPBs [17]. Still, many studies have reported that adequate levels of HL were associated with appropriate HPBs [38,39]. Additionally, older adults with adequate HL have been found to promote and encourage their behaviors regarding complex health issues, especially during the COVID-19 pandemic. Although adequate HL has been continuously associated with HPBs during the COVID-19 pandemic in various countries, studies in some contexts have shown the opposite findings, where older adults with adequate HL were not associated with COVID-19 preventive behaviors [20]. The effect of HL on HPBs among older adults with NCDs warrants further examination.

ACPM directly affected HPBs among older adults with NCDs in this study. This finding may be related to the fact that individuals with good accessibility to COVID-19 prevention materials are more likely to have better HPBs, especially during the COVID-19 pandemic. Our findings are consistent with Pechrapa and colleagues [40], who found that older adults in urban areas with good ACPM could access health information, health services, and COVID-19 preventive materials. To prevent and control the spread of COVID-19, the Thai government and healthcare agencies provided some COVID-19 prevention materials, including face masks and soap or alcohol for hand washing to residents of Thailand; however, some people were unable to access free COVID-19 prevention materials. Interestingly, most participants of this study had a monthly income of less than USD 143, which is lower than the average income in Thailand [41]. Thus, it could affect HPBs among older adults who do not have access to preventive materials for COVID-19, particularly older adults in rural communities with lower ACPM. During the new normal of the COVID-19 era, lifestyle and health behaviors changed due to pandemic control measures, significantly affecting older adults with NCDs, particularly people living in poverty [42]. However, health behaviors and disparities in access to healthcare need to be addressed to improve HPBs among all older adults with NCDs [15].

SN also had a positive direct effect on HPBs in our study, in which most of the participants had more social engagement. A previous study supports our findings that older adults who receive good support from family and friends are associated with good COVID-19 preventive behaviors (OR: 2.05, 95% CI: 1.10–3.82) [20]. Maintaining physical and mental well-being throughout life and into old age, SN contributes to HPBs and quality of life [43]. A recent integrative review of factors associated with HPBs during the COVID-19 pandemic revealed that more social engagement with SNs was associated with HPBs among older adults [44]. These findings have been supported with a prior SEM study that examined the factors affecting HPBs among older women in Korea [45]. Social support, including positive social interactions, was an influential factor with a direct effect on HPBs in older women [45]. These findings could explain why social support for older adults is essential for encouraging participation in social activities and maintaining positive relationships and interactions with families, relatives, and peers to establish SNs and improve mental well-being and quality of life [46]. Additionally, SNs are a positive resource that can be obtained through social interactions with family and friendships that enable an individual to live as fully as possible despite any current health conditions [47]. However, many studies revealed that an individual's lack of SN ties increased their risk of several diseases, morbidity, and mortality [48,49].

The main strength of the current study is its originality, as it was the first study to develop and test a hypothesized causal model of HPBs among older adults with NCDs who lived in Thai urban and rural communities during the new normal post-COVID-19 era. Participants' ages, education levels, current occupations, and monthly income were positively associated with HPBs. The findings also showed that older adults with NCDs living in urban communities had statistically significantly higher PSE and ACPM scores than older adults in rural communities. Furthermore, HPBs among older adults with NCDs were directly influenced positively by PSE, ACPM, HL, and SN, which could explain 81.0% of the variance in HPBs. These findings highlight significant factors influencing HPBs among urban and rural older adults with NCDs in this transition period of the new normal post-COVID-19 era that health policymakers and healthcare providers can apply to develop proper interventions and healthcare activities according to the local community's needs and cultural contexts.

There are some limitations to this study. First, the cross-sectional nature limits our ability to establish cause and effect of HPBs among urban and rural older adults with NCDs during the new normal post-COVID-19 era. Second, we collected data using self-reported questionnaires in older adults, and bias in this survey is possible. Third, we did not explore HPBs of older adults in different types of NCDs; future research should compare the different types of older adults with NCDs or focus on specific types of NCDs (e.g., hypertension, diabetes, or heart failure). Finally, the data were collected from only one province in Thailand due to budgetary limitations; thus, caution should be used when generalizing these results to other regions.

5. Conclusions

The current study showed that the PSE and ACPM of older adults with NCDs in urban communities were higher than in rural communities. The causal model of HPBs among these populations obtained a good fit with empirical data, which highlighted that PSE, ACPM, HL, and SN directly affect HPBs. Notably, healthcare providers should consider all significant factors to develop comprehensive interventions or healthcare activities for HPBs among older adults with NCDs according to their needs and cultural contexts. Attention should be paid to health behaviors and disparities in both urban and rural communities that affect HPBs among older adults with NCDs in order to encourage HPBs for as long as possible in old age.

Author Contributions: Conceptualization, W.S., S.T. and J.S.; methodology, W.S. and S.T.; software, W.S. and J.S.; validation, W.S., S.T. and J.S.; formal analysis, W.S.; investigation, W.S. and S.T.; writing—original draft preparation, W.S.; writing—review and editing, S.T. and J.S; supervision, S.T. and J.S. All authors have read and agreed to the published version of the manuscript.

Funding: This research received no external funding.

Institutional Review Board Statement: The study was conducted in accordance with the Declaration of Helsinki and approved by the Institutional Review Board of Krirk University (protocol code 0204/001 and dated 1 July 2022) and the Institutional Review Board of Ubon Ratchathani Provincial Health Office (protocol code SSJ.UB 2565-115 and dated 9 September 2022).

Informed Consent Statement: Informed consent was obtained from all subjects involved in the study.

Data Availability Statement: The data sets of this study are not publicly available due to the information that could compromise the research participants' privacy. However, data may be shared upon request to the authors.

Acknowledgments: We thank the participants of the study for their valuable contributions. This study was partially funded for publication by Krirk University, Bangkok, Thailand.

Conflicts of Interest: The authors declare no conflict of interest.

References

1. Huang, C.; Wang, Y.; Li, X.; Ren, L.; Zhao, J.; Hu, Y.; Zhang, L.; Fan, G.; Xu, J.; Gu, X.; et al. Clinical features of patients infected with 2019 novel coronavirus in Wuhan, China. *Lancet* **2020**, *395*, 497–506. [CrossRef] [PubMed]
2. Batista, C.; Shoham, S.; Ergonul, O.; Hotez, P.; Bottazzi, M.E.; Figueroa, J.P.; Gilbert, S.; Gursel, M.; Hassanain, M.; Kang, G.; et al. Urgent needs to accelerate the race for COVID-19 therapeutics. *EClinicalMedicine* **2021**, *36*, 100911. [CrossRef] [PubMed]
3. Bok, K.; Sitar, S.; Graham, B.S.; Mascola, J.R. Accelerated COVID-19 vaccine development: Milestones, lessons, and prospects. *Immunity* **2021**, *54*, 1636–1651. [CrossRef] [PubMed]
4. Khan, N.A.; Al-Thani, H.; El-Menyar, A. The emergence of new SARS-CoV-2 variant (Omicron) and increasing calls for COVID-19 vaccine boosters-The debate continues. *Travel Med. Infect. Dis.* **2022**, *45*, 102246. [CrossRef] [PubMed]
5. Bouzid, D.; Visseaux, B.; Kassasseya, C.; Daoud, A.; Fémy, F.; Hermand, C.; Truchot, J.; Beaune, S.; Javaud, N.; Peyrony, O.; et al. Comparison of Patients Infected with Delta Versus Omicron COVID-19 Variants Presenting to Paris Emergency Departments. *Ann. Intern. Med.* **2022**, *175*, 831–837. [CrossRef] [PubMed]
6. Centers for Disease Control and Prevention. WHO Coronavirus (COVID-19) Dashboard. Available online: https://covid19.who.int/ (accessed on 8 December 2022).
7. Ruengorn, C.; Awiphan, R.; Phosuya, C.; Ruanta, Y.; Wongpakaran, N.; Wongpakaran, T.; Thavorn, K.; Nochaiwong, S. Psychological Resilience and Adverse Mental Health Issues in the Thai Population during the Coronavirus Disease 2019 Pandemic. *Int. J. Environ. Res. Public Health* **2022**, *19*, 13023. [CrossRef]
8. Lu, G.; Zhang, Y.; Zhang, H.; Ai, J.; He, L.; Yuan, X.; Bao, S.; Chen, X.; Wang, H.; Cai, J.; et al. Geriatric risk and protective factors for serious COVID-19 outcomes among older adults in Shanghai Omicron wave. *Emerg. Microbes Infect.* **2022**, *11*, 2045–2054. [CrossRef]
9. Centers for Disease Control and Prevention. People with Certain Medical Conditions. Available online: https://www.cdc.gov/coronavirus/2019-ncov/need-extra-precautions/people-with-medical-conditions.html (accessed on 8 December 2022).
10. Kitchanapaibul, S.; Udplong, A.; Apidechkul, T.; Tamornpark, R.; Mulikaburt, T.; Srichan, P.; Suratana, S.; Yeemard, F.; Wongnuch, P. Experiences and expectations regarding COVID-19 prevention and control measures among the hill tribe population of northern Thailand: A qualitative study. *BMC Public Health* **2021**, *21*, 1060. [CrossRef]
11. Daly, J.R.; Depp, C.; Graham, S.A.; Jeste, D.V.; Kim, H.-C.; Lee, E.E.; Nebeker, C. Health Impacts of the Stay-at-Home Order on Community-Dwelling Older Adults and How Technologies May Help: Focus Group Study. *JMIR Aging* **2021**, *4*, e25779. [CrossRef]
12. Chantakeeree, C.; Sormunen, M.; Jullamate, P.; Turunen, H. Health-promoting Behaviors among Urban and Rural Older Thai Adults with Hypertension: A Cross-sectional Study. *Pac. Rim Int. J. Nurs. Res.* **2021**, *25*, 242–254.
13. Pender, N.J.; Murdaugh, C.L.; Parsons, M.A. *Health Promotion in Nursing Practice*, 7th ed.; Pearson: Columbia, SC, USA, 2015.
14. Zhu, D.; Ozaki, A.; Virani, S.S. Disease-Specific Excess Mortality during the COVID-19 Pandemic: An Analysis of Weekly US Death Data for 2020. *Am. J. Public Health* **2021**, *111*, 1518–1522. [CrossRef] [PubMed]
15. Pothisiri, W.; Vicerra, P.M.M.; Buathong, T. Poverty, noncommunicable diseases, and perceived health risks among older adults during the COVID-19 pandemic in urban Thailand. *Asian Soc. Work. Policy Rev.* **2022**, *16*, 126–135. [CrossRef] [PubMed]
16. Zhu, W.; Cheng, Z.; Howard, V.J.; Judd, S.E.; Blair, S.N.; Sun, Y.; Hooker, S.P. Is adiposity associated with objectively measured physical activity and sedentary behaviors in older adults? *BMC Geriatr.* **2020**, *20*, 257. [CrossRef] [PubMed]
17. Yusefi, A.R.; Barfar, E.; Daneshi, S.; Bayati, M.; Mehralian, G.; Bastani, P. Health literacy and health promoting behaviors among inpatient women during COVID-19 pandemic. *BMC Women's Health* **2022**, *22*, 77. [CrossRef] [PubMed]
18. Wungrath, J.; Autorn, N. Factors Influencing Health Promotion Behaviors and Prevention of COVID-19 among Elderly during the Second Wave of Pandemic in Chiang Mai, Thailand. *Ann. Rom. Soc. Cell Biol.* **2021**, *25*, 1702–1715.
19. Posai, V.; Suksatan, W.; Choompunuch, B.; Koontalay, A.; Ounprasertsuk, J.; Sadang, J.M. Assessment of the Health-Promoting Behaviors of Hospitalized Patients with Non-Communicable Diseases during the Second Wave of COVID-19. *J. Multidiscip. Healthc.* **2021**, *14*, 2185–2194. [CrossRef]

20. Yodmai, K.; Pechrapa, K.; Kittipichai, W.; Charupoonpol, P.; Suksatan, W. Factors Associated with Good COVID-19 Preventive Behaviors among Older Adults in Urban Communities in Thailand. *J. Prim. Care Community Health* **2021**, *12*, 21501327211036251. [CrossRef]
21. Lee, M.K.; Oh, J. Health-Related Quality of Life in Older Adults: Its Association with Health Literacy, Self-Efficacy, Social Support, and Health-Promoting Behavior. *Healthcare* **2020**, *8*, 407. [CrossRef]
22. Foundation of Thai Gerontology Research and Development Institute. *Situation of the Thai Older Persons 2020*; Institute for Population and Social Research, Mahidol University: Nakhon Pathom, Thailand, 2021.
23. National Statistical Office. Reporting to a Study of Guidelines for Defining Urban and Rural Areas in Thailand. Available online: http://statstd.nso.go.th/download.aspx (accessed on 6 December 2022).
24. Jackson, D.L. Sample Size and Number of Parameter Estimates in Maximum Likelihood Confirmatory Factor Analysis: A Monte Carlo Investigation. *Struct. Equ. Model. A Multidiscip. J.* **2001**, *8*, 205–223. [CrossRef]
25. Faul, F.; Erdfelder, E.; Buchner, A.; Lang, A.-G. Statistical power analyses using G*Power 3.1: Tests for correlation and regression analyses. *Behav. Res. Methods* **2009**, *41*, 1149–1160. [CrossRef]
26. Cohen, J. *Statistical Power Analysis for the Behavioral Sciences*; L. Erlbaum Associates: Hillsdale, NJ, USA, 1988.
27. Health Education Division. Health Literacy and Health-Promoting Behavior Assessment Tools for the Village to Change Health Behavior. Available online: http://www.hed.go.th/linkHed/424 (accessed on 10 December 2021).
28. Lubben, J.; Blozik, E.; Gillmann, G.; Iliffe, S.; von Renteln Kruse, W.; Beck, J.C.; Stuck, A.E. Performance of an Abbreviated Version of the Lubben Social Network Scale among Three European Community-Dwelling Older Adult Populations. *Gerontologist* **2006**, *46*, 503–513. [CrossRef] [PubMed]
29. Becker, H.; Stuifbergen, A.; Oh, H.S.; Hall, S. Self-rated abilities for health practices: A health self-efficacy measure. *J. Health Behav. Educ. Promot.* **1993**, *17*, 42–50.
30. Kline, R.B. *Principles and Practice of Structural Equation Modeling*, 2nd ed.; Guilford Publications: New York, NY, USA, 2005.
31. Giena, V.P.; Thongpat, S.; Nitirat, P. Predictors of health-promoting behaviour among older adults with hypertension in Indonesia. *Int. J. Nurs. Sci.* **2018**, *5*, 201–205. [CrossRef] [PubMed]
32. Mszar, R.; Buscher, S.; McCann, D.; Taylor, H.L. Self-Efficacy, Perceived Barriers to Care, and Health-Promoting Behaviors among Franco-Americans across Cardiovascular Risk Factors: A Cross-Sectional Study. *Am. J. Health Promot.* **2020**, *35*, 703–707. [CrossRef] [PubMed]
33. Bandura, A. *New Developments in Goal Setting and Task Performance*; Routledge: New York, NY, USA, 2013.
34. Tan, F.C.J.H.; Oka, P.; Dambha-Miller, H.; Tan, N.C. The association between self-efficacy and self-care in essential hypertension: A systematic review. *BMC Fam. Pract.* **2021**, *22*, 44. [CrossRef] [PubMed]
35. DDo, B.N.; Nguyen, P.-A.; Pham, K.M.; Nguyen, H.C.; Nguyen, M.H.; Tran, C.Q.; Nguyen, T.T.P.; Tran, T.V.; Pham, L.V.; Tran, K.V.; et al. Determinants of Health Literacy and Its Associations with Health-Related Behaviors, Depression among the Older People with and without Suspected COVID-19 Symptoms: A Multi-Institutional Study. *Front. Public Health* **2020**, *8*, 581746. [CrossRef] [PubMed]
36. Nutbeam, D. Health literacy as a public health goal: A challenge for contemporary health education and communication strategies into the 21st century. *Health Promot. Int.* **2000**, *15*, 259–267. [CrossRef]
37. Nutbeam, D. Defining and measuring health literacy: What can we learn from literacy studies? *Int. J. Public Health* **2009**, *54*, 303. [CrossRef]
38. Kim, S.; Song, Y.; Park, J.; Utz, S. Patients' Experiences of Diabetes Self-Management Education according to Health-Literacy Levels. *Clin. Nurs. Res.* **2020**, *29*, 285–292. [CrossRef]
39. Ueno, H.; Ishikawa, H.; Suzuki, R.; Izumida, Y.; Ohashi, Y.; Yamauchi, T.; Kadowaki, T.; Kiuchi, T. The association between health literacy levels and patient-reported outcomes in Japanese type 2 diabetic patients. *SAGE Open Med.* **2019**, *7*, 2050312119865647. [CrossRef] [PubMed]
40. Pechrapa, K.; Yodmai, K.; Kittipichai, W.; Charupoonpol, P.; Suksatan, W. Health Literacy among Older Adults during COVID-19 Pandemic: A Cross-Sectional Study in an Urban Community in Thailand. *Ann. Geriatr. Med. Res.* **2021**, *25*, 309–317. [CrossRef] [PubMed]
41. National Statistical Office of Thailand. Summary Survey of Household Socio-Economic Conditions in the First 6 Months of 2021. Available online: http://www.nso.go.th/sites/2014en/Survey/social/household/household/2021/pocketbookSES_64.pdf (accessed on 7 December 2022).
42. Bukhman, G.; Mocumbi, A.O.; Atun, R.; Becker, A.E.; Bhutta, Z.; Binagwaho, A.; Clinton, C.; Coates, M.M.; Dain, K.; Ezzati, M.; et al. The Lancet NCDI Poverty Commission: Bridging a gap in universal health coverage for the poorest billion. *Lancet* **2020**, *396*, 991–1044. [CrossRef] [PubMed]
43. Pizzi, M.A.; Richards, L.G. Promoting Health, Well-Being, and Quality of Life in Occupational Therapy: A Commitment to a Paradigm Shift for the Next 100 Years. *Am. J. Occup. Ther.* **2017**, *71*, 7104170010p1–7104170010p5. [CrossRef] [PubMed]
44. Suksatan, W.; Teravecharoenchai, S.; Sarayuthpitak, J. Factors Associated with a Health-promoting Lifestyle among Adults and Older Adults in the Era of COVID-19: An Integrative Review. *Open Access Maced. J. Med. Sci.* **2022**, *10*, 725–732. [CrossRef]
45. Oh, J. Factors affecting health promoting behavior among older women in Korea: A structural equation model. *Health Promot. Int.* **2021**, *36*, 924–932. [CrossRef]
46. Allahverdipour, H.; Karimzadeh, Z.; Alizadeh, N.; Asghari Jafarabadi, M.; Javadivala, Z. Psychological well-being and happiness among Middle-aged women: A cross-sectional study. *Health Care Women Int.* **2021**, *42*, 28–42. [CrossRef]

47. Lubben, J.E. Assessing social networks among elderly populations. *Fam. Community Health* **1988**, *11*, 42–52. [CrossRef]
48. Chang, Q.; Chan, C.H.; Yip, P.S.F. A meta-analytic review on social relationships and suicidal ideation among older adults. *Soc. Sci. Med.* **2017**, *191*, 65–76. [CrossRef]
49. Santini, Z.I.; Koyanagi, A.; Tyrovolas, S.; Haro, J.M. The association of relationship quality and social networks with depression, anxiety, and suicidal ideation among older married adults: Findings from a cross-sectional analysis of the Irish Longitudinal Study on Ageing (TILDA). *J. Affect. Disord.* **2015**, *179*, 134–141. [CrossRef]

nutrients

MDPI

Article

Lifestyle Intervention Randomized Controlled Trial for Age-Related Macular Degeneration (AMD-Life): Study Design

Alexandra P. M. de Koning-Backus [1,2], Jessica C. Kiefte-de Jong [2,3], Jeroen G. J. van Rooij [2,4], AMD-Life Team [5,†], André G. Uitterlinden [2,4], Trudy G. Voortman [2,6], Magda A. Meester-Smoor [1,2,‡] and Caroline C. W. Klaver [1,2,7,8,*,‡]

1 Department of Ophthalmology, Erasmus MC University Medical Center Rotterdam, P.O. Box 2040, 3000 CA Rotterdam, The Netherlands
2 Department of Epidemiology, Erasmus MC University Medical Center Rotterdam, P.O. Box 2040, 3000 CA Rotterdam, The Netherlands
3 Department of Public Health and Primary Care, Health Campus The Hague, Leiden University Medical Center, 2333 ZA The Hague, The Netherlands
4 Department of Internal Medicine, Erasmus MC University Medical Center Rotterdam, P.O. Box 2040, 3000 CA Rotterdam, The Netherlands
5 AMD-Life Team, Erasmus MC University Medical Center Rotterdam, P.O. Box 2040, 3000 CA Rotterdam, The Netherlands
6 Division of Human Nutrition and Health, Wageningen University & Research, P.O. Box 17, 6700 AA Wageningen, The Netherlands
7 Department of Ophthalmology, Radboud University Medical Center, P.O. Box 9101, 6500 HB Nijmegen, The Netherlands
8 Institute of Molecular and Clinical Ophthalmology, University of Basel, CH-4031 Basel, Switzerland
* Correspondence: c.c.w.klaver@erasmusmc.nl
† The AMD-Life team, members are summarized in the supplementary material.
‡ These authors contributed equally to this work.

Abstract: Age-related macular degeneration (AMD) has a strong genetic basis, but environmental factors such as smoking and a healthy diet can decrease the genetic fate by up to 50%. Current guidelines for clinical management include recommendations for a healthy lifestyle and antioxidant supplementation. However, many ophthalmologists do not inform their patients of this AMD-beneficial lifestyle. An important reason is the lack of trust that transition of lifestyle will be feasible in persons of advanced age and lack of methodology to measure lifestyle or its biological effects. To address these issues, we set up the lifestyle intervention study AMD-Life. It aims to investigate whether personalized risk-profiling (including genetic testing) and/or additional coaching can motivate patients to change their lifestyle. It also explores which biomarkers best reflect lifestyle change beneficial for AMD. The first year is a three-arm, self-contained open-label randomized clinical trial. A total of 150 AMD patients aged 55–85 years were randomized into three arms: (A) merely standard recommendations; (B) A conditions plus personalized risk profiling based on genetics and lifestyle, (C) B conditions plus coaching. The second year tests sustainability of lifestyle changes without active intervention. AMD-Life can provide further insight into the relevance of these interventions for the clinical management of AMD.

Keywords: age-related macular degeneration; lifestyle; nutrients; Mediterranean diet; genetic testing; personalized risk-profiling; gut microbiome; behavioral change technique; coaching

1. Introduction

Age-related macular degeneration (AMD) is a leading cause of blindness in the elderly of the Western world. Currently, more than 196 million people worldwide have AMD, and global aging will increase the number of affected persons to 288 million by the year 2040 [1,2]. The prevalence of early and intermediate AMD is ~20% in people aged over 65 years. These stages are characterized by the presence of yellow-white deposits (drusen) under the retinal pigment epithelium (RPE) and/or pigment changes, usually occurring without visual implications. The prevalence of the vision-disabling late AMD ranges from 3% in individuals aged 65 years to ~13% in those over 90 years [3]. This stage is characterized by either macular neovascularization (MNV, neovascular or wet AMD) or loss of RPE and photoreceptors (geographic atrophy, GA, or dry AMD) in the macular area. The transition from intermediate to the late stage of the disease occurs with a frequency of 3–6% per year [4]. Within 10 years, this will increase up to ~60%. The current therapeutic options for wet AMD are anti-VEGF injections, which improve vision drastically in the short term but often fail to protect against severe vision loss in the long term [5]. Therapy for GA has recently made progress with regimens controlling the complement pathway, but this still needs confirmation in real-world studies [6,7].

The genetic architecture of AMD has been mostly uncovered. Several years ago, a large genome-wide association study reported association with 52 common genetic variants in 34 loci [7]. The variants represented prominent pathways such as the complement cascade, lipid homeostasis, and extra-cellular matrix. A genetic risk score (GRS) that aggregated risks of these common variants showed a discriminative accuracy of 0.84 to correctly identify a person with late AMD [8]. In addition to common variants, many rare variants have been identified in genes involved in complements [9], collagen [10], coagulation pathways, and metalloproteinases [11]. Remarkably, although AMD is now one of the best genetically characterized complex diseases [7], genetic testing for AMD is still mostly a scientific enterprise and not a routine assessment in clinical practice. An important reason for this is the position of ophthalmic professional organizations, which do not advise genetic testing for AMD until treatment for specific disease-associated genotypes has been established.

Environmental factors also play a significant role in disease pathogenesis; in particular, smoking [12–15], dietary patterns [16–18], and physical activity [19] have been associated. We recently showed that persons in the most unfavorable tertile of these lifestyle variables had an increased risk of late AMD by at least two-fold [8]. Reduction of the risk is possible with nutrients that have antioxidative properties and polyunsaturated omega-3 fatty acids [16,20,21]. International dietary guidelines (United States, Canada, Germany, France, United Kingdom, The Netherlands, Japan, e.g.) recommend less meat and more plant-based dietary patterns including vegetables, fruit, whole grains, legumes, nuts and seeds, and (fatty) fish [22]. These recommendations are comparable with the Mediterranean diet pattern [23,24] and are also in line with the advice of the UN Food and Agriculture Organization and the World Health Organization [22,25]. However, data show that adherence to these recommendations is low [16,26,27]. In the Rotterdam Study, we recently investigated recommended minimum intake values (RMIV) of vegetables, fruit, and fish and the risk of AMD [16]. We found that only 4% of the participants consumed the recommended amounts of foods. However, those who did showed a significant decreased hazard ratio by more than half of incidence of advanced AMD. Two large clinical trials (AREDS 1; AREDS 2) provided evidence that supplementation with antioxidants and zinc can also significantly reduce disease progression [28,29]. Novel is the finding that intestinal microbiota [30] are associated with AMD, which may mediate the association with dietary factors [5,17,30], oral supplements [29], and other environmental factors [29].

Most preferred practice patterns for AMD recommend lifestyle measures, but the information provided to patients and adherence tend to be poor. Up to 60% of patients do not take oral supplements or use them incorrectly, and smoking cessation or long-term dietary changes are often not achieved [31–34]. One reason may be that retinal specialists do not convey the message adequately due to limited time, knowledge, or doubts about effectiveness [33–37]. This refutes the current view that healthcare professionals should play a significant role in motivating patients to change habitual patterns or behaviors that pose significant health risks. How does one motivate patients to take action and change behavior? A recent insight is that, at certain times, patients appear to be more receptive to behavioral change and lifestyle advice than usual [36,37], for example, when a disease has been diagnosed, they have received abnormal test results, or have been confronted with a high genetic risk. These are called 'teachable moments' and may be considered as optimal timing to start lifestyle counselling [36,37].

In the current AMD-Life trial, we aim to investigate which strategy is most successful in motivating patients to change the behavior that is driving them to vision loss: receiving only general information on risks and benefits, receiving detailed information about personalized risks and targets for change, or receiving coaching based on behavioral change technique (BCT) [38,39] and motivational interviewing (MI) [40,41]. We also aim to explore whether biomarkers in blood and the gut microbiome can be used to monitor lifestyle and lifestyle change, which can then be used to motivate patients further. This article provides the rationale and design of the study.

2. Materials and Methods

2.1. Study Design and Population

AMD-Life is a 2-year lifestyle intervention trial. The first year is a self-contained open-label randomized clinical trial with 3 intervention arms. The second year aims to investigate the sustainability of improved lifestyle without active intervention. Patients visit the research center 3 times: at baseline, month 12, and month 24. Ethical approval was obtained from the Medical Ethical Committee at Erasmus MC (MEC-2018-063). AMD-Life is registered at ClinicalTrials.gov (registration identifier: NCT05667441).

This study is designed as an inception phase study. We set up the main elements, test our assumptions and feasibility, check procedures, and focus on adherence, reporting participant and physician experience for the three different strategies tested, specifically targeting long-term behavior change.

Sample size calculation indicated that 50 eligible subjects per group would be sufficient to detect an effect size of 0.25 for adherence to lifestyle recommendations after the 12-month intervention (alpha: 0.05, power: (1-beta err prob.): 0.8897, calculated by G*Power).

Recruitment for the study started 2 August 2021 and is expected to end 1 July 2023. Participants are made aware of the study by their attending physician at outpatient clinics (OPC) of Erasmus MC Rotterdam, the Rotterdam Eye Hospital, Radboudumc, in Nijmegen, and other hospitals in the Netherlands. Interested candidates can apply online through the website: www.maculadegeneratie.nl (accessed on January 2023) with the apply button for participation, or by email or telephone.

Inclusion criteria: age between 55 years and 85 years, a diagnosis of intermediate AMD or unilateral late AMD, as determined by multimodal imaging and approved by the EyeNED Reading Center.

Exclusion criteria: participation in other intervention studies for AMD; living in retirement homes (because of difficulty in implementation of diet); dementia; macular pathology other than AMD; liver and kidney insufficiency; persons who are illiterate and have no independent trusted person with them to explain the informed consent form.

Prior to the study, all potential participants are informed and asked to provide written consent. Subsequently, participants are invited to visit the research center. Eligible participants are randomly assigned using a block-wise randomization procedure (in a 1:1:1 ratio) to one of three study arms. The assignment is by means of a computer-generated table with a random number sequence. This table gives a random order of allocation to the 'Treatment Group'. The order is by date and time of the first study visit and stratified by gender and age categories (55–64, 65–74, 75–85). The first 125 participants are randomized per block (sets) of 6, and the remaining 25 per sets of 3. The researchers are blinded for randomization.

2.2. Interventions

The AMD-Life study tests three study arms, which each represent a different strategy to stimulate and increase adherence to a healthy lifestyle: (A) patients receive standard recommendations and supplements free of charge; (B) as in A, but patients additionally receive personal risk profiling based on phenotype as well as genetic and environmental risk factors; (C) as in B, but patients additionally receive professional coaching. The schematic overview of inclusion, exclusion, the three arms, and the follow-up is shown in Figure 1.

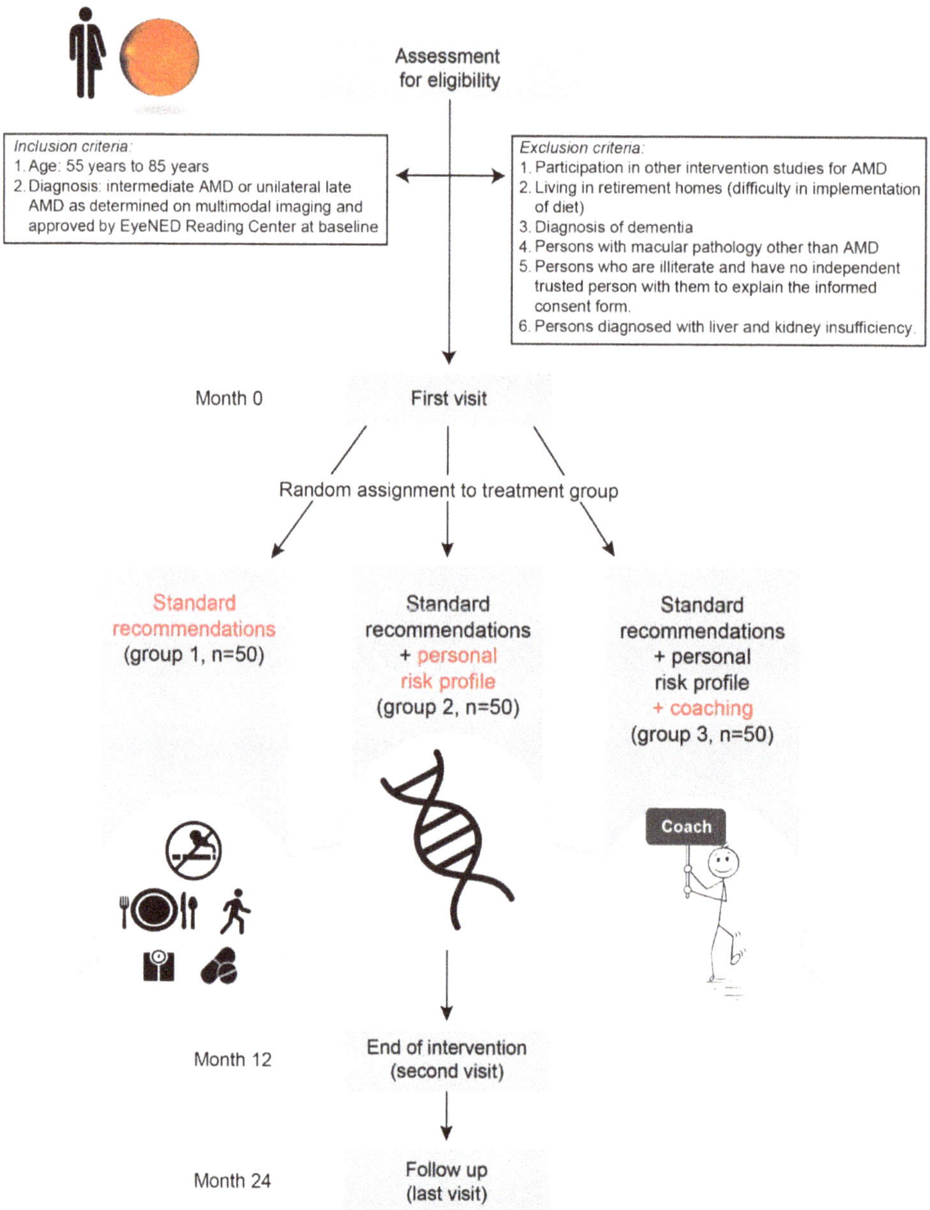

Figure 1. Flow-chart of the AMD-Life study design. Inclusion and follow-up of study participants in each study arm. Three study arms test different strategies to stimulate and increase adherence to a healthy lifestyle. After 12 months, the study arms end, and a single follow-up moment at 24 months is scheduled for assessment of sustainability of lifestyle change.

2.2.1. Group A

Standard recommendations plus AREDS 2 and Omega 3 supplementation ('standard care')—n = 50. At the baseline visit, an experienced trial nurse informs the participant of personal outcomes of the measurements performed at the research center and of the standard AMD lifestyle recommendations. Participants also receive this information in a letter. The recommendations include refrain from smoking, perform regular physical exercise, strive for BMI < 25, and increase the intake of the dietary food groups of (green leafy) vegetables, fruit, (fatty) fish, and legumes. The Mediterranean Diet pattern is used as guidance for recommendations (for a more detailed description see Table 1). Participants receive oral supplements with antioxidants according to the AREDS 2 formula and Omega 3 forte fish oil free of charge (Table S1).

Table 1. Mediterranean diet score instrument (MeDiet).

	Questions	Criteria for 0.5 point	Criteria for 1 Point	Included in LF Scoring
1	Do you use olive oil as the main culinary fat?		Yes	√
2	How much olive oil do you consume per day (incl. oil used for frying, salads, out-of-house meals, etc.)?	2–3 tbsp	≥4 tbsp	√
3	How many grams of vegetable (incl. raw or as a salad) do you consume per day?	200–400 g	≥400 g	√
4	How many fruit units (incl. natural fruit juices) do you consume per day? (units of 80 g each)	2	≥3	√
5	How often do you eat meat?		<1 p/d	
6	If you consume meat, what kind of meat do you consume most of the time? Do you preferentially consume lean meat: chicken, turkey, or rabbit meat, instead of red meat: veal, pork, hamburger, or sausage?	Both equally often	lean meat	√
7	How many servings of meat do you consume per week?		<100–150 g	√
8	How many servings of butter, margarine, or cream do you consume per day? (1 serving: 12 g)		<12 g	√
9	How many sweetened and/or carbonated beverages do you drink per day?		<1 gl	√
10	How many glasses of red wine do you consume per week? (1 glass = 124 mL)		≤3 gl	
11	How many grams of legumes do you consume per week? (450 g per week = 65 g per day)	300–450 g	≥450 g	√
12	How many servings of fish or shellfish do you consume per week? (1 fish = 100–150 g or 4–5 units of shellfish = 200 g)	150–300 g	≥300 g	√

Table 1. *Cont.*

	Questions	Criteria for 0.5 point	Criteria for 1 Point	Included in LF Scoring
13	How often per week do you consume fatty fish? (Salmon, tuna, herring, eel, sardines, mackerel)		1 p/w	
14	How many times a week do you eat sweets or pastries such as cakes, cookies, and biscuits?		<3	√
15	How many servings of nuts (including peanuts) do you consume per week? (1 serving = 30 g = about a handful)	30–89 g	≥90 g	√
16	How many times per week do you consume vegetables, pasta, rice, or other dishes seasoned with sofrito (sauce made with tomato and onion, leek, or garlic and simmered with olive oil)?	1	≥2	√

Short questionnaire to assess adherence to the 16-point MeDiet. Abbreviations: LF, Lifestyle Form; incl., including; etc., et cetera; tbsp, tablespoon; g, gram; p/w, per week; p/d, per day; gl, glass.

2.2.2. Group B

Standard care plus personalized risk profiling—n = 50. Participants receive a personalized risk profile for motivational purposes. They are informed about their AMD genetic risk (low, medium, high, very high) based on the GRS, risk of progression, and the potential gain from lifestyle improvement. The four lifestyle factors of smoking, BMI, diet, and physical exercise are graded using a 13-point easy-to-use lifestyle scoring form (Figure S1 - page 1); participants receive their personal grades and can check their room for improvement.

2.2.3. Group C

Standard care plus personal risk profiling plus coaching—n = 50. Participants receive coaching based on MI and the stages of BCTs using the AMD-Life coaching model (Figure 2): determination of awareness, motivation, perseverance and discipline, and goal setting. Depending on the stage and/or need of the participant, several tools, information leaflets, documents with assignments, and tables (e.g., specific foods and levels/concentrations of beneficial AMD-nutrients) are provided to the participant to enhance awareness and stimulate habitual change. Coaching sessions take place online by a 15–20 min videoconference twice a month for the first 2 months, followed by once a month for the following 10 months.

After an intervention of 12 months, all study participants visit the research center, after which the participants form a single group to observe the sustainability of lifestyle improvements. At this point, Group A receives insights into their lifestyle changes over the first year and is informed about their genetic risk, similarly to groups B and C, who received this information at month 0 (Figures S1 and S2). At month 24, the study ends with a third visit to the research center.

2.3. *Avoiding Attrition Bias*

Multiple strategies are employed to reduce attrition in all three arms, including an effective recruitment process, brief and informative encounters with staff members, a good tracking system to facilitate participant contact, anti-oxidant and fish oil supplementation free of charge, and the development of a trusting and collaborative relationship between researchers and participants.

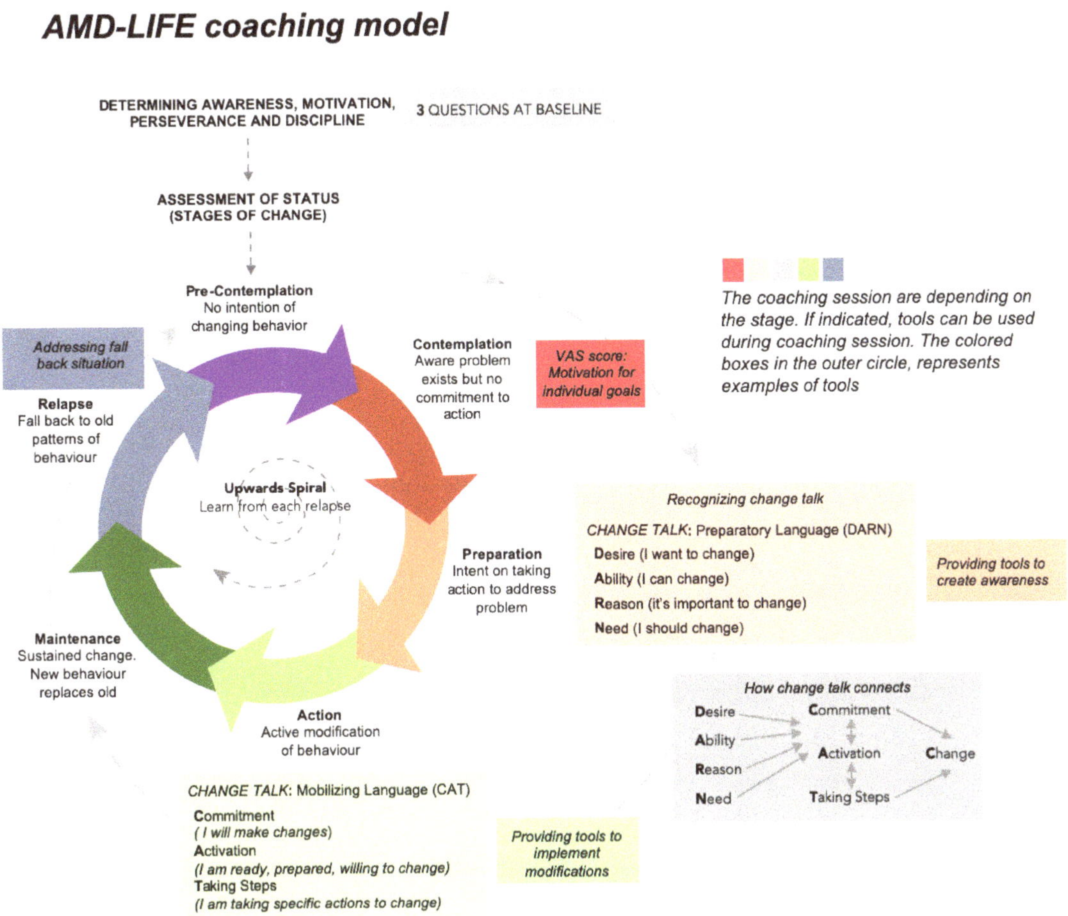

Figure 2. AMD-Life coaching model. The AMD-Life coaching model reflects an upwards spiral to learn from each relapse. Included in the model are: 'the 3 questions' at baseline to determine awareness, motivation, perseverance and discipline, and goal-setting; a model for Behavioral Change Technique (BCT), the transtheoretical model of change (determining of ambivalence/assessment of status), adapted with permission from Prochaska and DiClemente [39,41]; and a model for motivational interviewing (MI), increasing/recognizing of 'change talk' through DARN-CAT, adapted with permission from Miller and Rollnick [40]. Depending on the stage and/or need of the participant, several tools, information leaflets, documents with assignments, and tables (e.g., specific foods and levels/concentrations of beneficial AMD-nutrients) can be used and/or provided to enhance awareness and stimulate habitual change.

If a participant wishes to withdraw from the study intervention, a telephone call is made prior to withdrawal to discuss the reason and possible solutions. If there is no solution, the reason for withdrawal is documented in the participant records for the subsequent analysis and the interpretation of the results.

2.4. Data Collection and Measurement

Data collection is performed at month (mo.) 0, 3, 6, 9, 12, and 24. Outcome variables, data collection, measurements, and time points are depicted in Figure 3.

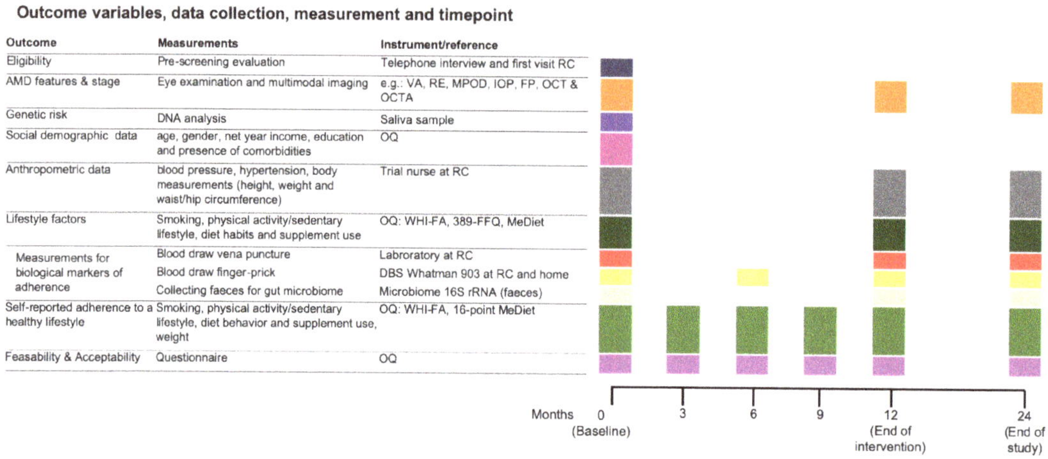

Figure 3. Data collection in AMD-Life. All study participants have the same measurements, irrespective of the study arm. Abbreviations: RC, research center; AMD, age-related macular degeneration; VA, best-corrected visual acuity; RE, refractive error; MPOD, Macular Pigment Optical Density; IOP, intra-ocular pressure; FP, fundus photography; OCT, optical coherence tomography; OQ, online questionnaire; WHI-FA, Women's Health Initiative Physical Activity Questionnaire; 389-FFQ, food frequency questionnaire; MeDiet, Mediterranean Diet Score; BMI, body mass index; DBS, dried blood spot.

2.4.1. Ophthalmologic Examination

At mos. 0, 12, and 24, participants undergo a comprehensive ophthalmologic examination consisting of measurement of best-corrected visual acuity (VA); refractive error (TopCon TRK-2P objective refractometer); Macular Pigment Optical Density (MPOD; MPS II); intra-ocular pressure (IOP; non-contact tonometer); fundus photography, OCT, and OCTA (3D macula, 3D optic disk, wide, and angiography with T Topcon Triton plus, Topcon 3D OCT 2000-plus; Topcon TRC 50EX with a NIKON D7100 camera).

Features and stages of AMD are classified by grading of multimodal images according to a modification of WARMGS (CFP) [42] and IN-OCT (OCT) consensus by 4 experienced human graders. Quantification of lesions are determined using Deep Learning algorithms developed and available at EyeNED Reading Center [42–45].

2.4.2. Genetic Factors

Subjects provide a saliva sample for DNA analysis 1 month prior to visiting the research center. Genotyping is performed for known common AMD risk variants, rare variants associated with AMD, and mutations in AMD-mimicking genes. The genotyping platform used is the Illumina Global Screening version 3 MD array with a custom content containing >3500 probes for detecting rare variants and mutations (GSAv3MD-GOALLv1) in CFH (n = 198), CFI (n = 90), C3 (n = 71), C9 (n = 20), ABCA4 (n = 1239), TIMP3 (n = 9), PRPH2 (n = 71), BEST1 (n = 27), CTNNA1 (n = 37), and MFRP (n = 2). Total genetic risk score (GRS) for common variants is calculated by the following formula for GRS, where Gi represents the genotype of variant i and βi represents the effect size of variant i (natural logarithm of the fully conditioned odds ratio of the minor allele of variant i), based on the GWAS of the International Age-Related Macular Degeneration Genomics Consortium [7]. Genotypes were coded as 0, 1, or 2 or, if the genotype is available from imputation, based on the number of minor alleles. Detailed information on SNPs used in GRS calculation is available at (Table S2). GRS is subsequently stratified into tertiles based on the distribution of GRS in the EYE-RISK cohort as Low (GRS < −0.057), Intermediate (GRS > −0.057 &

< 1.131), or High (GRS > 1.1131) genetic risk [8]. High GRS is then stratified in equal proportion into High and Very High (GRS $\geq$ 3).

2.4.3. Covariates

All questionnaires used in the AMD-Life study are digital/online. The participants receive an e-mail with a secure link to the online questionnaires, and data are entered directly into the AMD-Life database.

Social Demographics

We ask about baseline social demographic factors by means of a short questionnaire at mo. 0. This includes gender, age, education, eye disorders, family history of AMD, occupation, presence of comorbidity (such as diabetes, cardiovascular disease, hypertension), and living area (6-digit postal code).

Anthropometric Data

Anthropometric data are measured and collected at the research center at mos. 0, 12, and 24. Body measurements, including height, weight, and waist and hip circumference, are measured in kilograms and centimeters. Weight and height are used to estimate the body mass index (BMI = weight/(height)2). Blood pressure, systolic and diastolic, are measured, using an automatic sphygmomanometer. Hypertension is defined as a systolic blood pressure $\geq$ 140 mmHg or a diastolic blood pressure $\geq$ 90 mmHg.

2.4.4. Biochemical Measurements

To explore whether biological markers in body specimen are associated with lifestyle change and may be used as a guide star, we draw blood by vena puncture (mos. 0, 12, 24) and by finger-prick (mos. 0, 6, 12, 24), and ask the participant to collect feces (mos. 0, 12, 24).

Vena Puncture

Biochemical measurements such as anti-oxidants, fat soluble vitamins, minerals, fatty acids, lipids, complement factors, and inflammatory markers are sampled from serum, EDTA plasma, and whole blood obtained by vena puncture. All samples are prepared for each required analysis according to standard methods. Aliquots are stored (−80 °C) for later analysis.

Finger-Prick

To assess whether results from finger prick measurements performed at home may also be used as biological markers of adherence, we ask participants to collect dried whole blood spots (DBS) on a 903 protein saver card from Whatman (mo. 6 at home, mos. 0, 12, 24 at RC) [46,47]. Finger-prick biomarkers assays include anti-oxidants, fat soluble vitamins, minerals, and fatty acids.

Collecting Feces

To measure diversity and composition (i.e., alpha and beta diversity and single taxa) of the fecal microbiome over time, we ask participants to collect a sample of their feces at home at three time points: mos. 0, 12, and 24. Participants receive a feces collection kit with all necessary materials. The sample is collected using a preservative and stored at –80 °C upon arrival at the research center for later analysis. Medication use (such as antibiotics) is registered as well (yes/no; if yes, which).

2.4.5. Lifestyle Factors

Smoking

Smoking habits are assessed by questionnaire at mos. 0, 3, 6, 9, 12, and 24. Questions include smoking of cigarettes, cigars, tobacco pipe: yes, currently/yes, but quit/no never/;

year of smoking cessation; number of cigarettes/cigars/pipe per day; number of years of smoking.

Dietary Assessment and Supplement Use

A 389-item food frequency questionnaire (389-FFQ) developed for Dutch adults [48,49] is used to measure dietary behavior at mos. 0, 12, and 24. The 389-FFQ records frequency of each intake as times per day, per week, or per month. The portion sizes are in grams per day (g/day). Nutrient intake data are calculated using the Dutch Food Composition Tables [50], in close collaboration with the Department of Human Nutrition, Wageningen University, the Netherlands. The 389-FFQ was validated against other dietary assessment methods, and adequate ranking of participants according to food and nutrient intake was demonstrated [48,49,51,52]. Oral supplement use (yes or no) is measured using self-reported questionnaires at different time points.

Adherence to Mediterranean diet is scored in two ways: (1) analysis of data from the 389-FFQ; (2) analysis of data from the validated 14-point MeDiet screening instrument [53,54] at mos. 0, 3, 6, 9, 12, and 24. The 14-point MeDiet screening instrument is a modified version of the original standardized Mediterranean Diet Score (MDS) [23,24]. We added two additional questions to this instrument in our version (Table 1).

The 16-point MeDiet is used as part of the lifestyle form that is available for groups B and C. In the lifestyle form, outcomes of 13 out of the 16 questions, as indicated in Table 1, are used.

Physical Activity

Physical exercise is assessed using the validated 'Women's Health Initiative (WHI) Physical Activity Questionnaire' [55] at mos. 0, 12, and 24. The WHI Physical Activity Questionnaire is a common instrument and has previously been used in several AMD studies [19]. The questionnaire contains questions on walking, cycling, gardening, various sports, and housekeeping, according to time spent in light, moderate, and vigorous activity.

Feasibility and Acceptability

To assess feasibility and acceptability, we ask participants about their own experience with AMD, their experience with the recommendations, and depending on the intervention group, we ask about their experience with the personalized risk score and additional coaching (mos. 3, 6, 9, 12, and 24) by a short questionnaire.

Assessment of Change in Habitual Behavior

All coaching sessions (Group C) are recorded by video and registered in the AMD-Life database. Reporting includes motivation, goal-setting for change, self-confidence according to the AMD-Life coaching model (which aims to reflect an upwards spiral), and learning from potential relapses (Figure 2). In order to start the coaching process, we ask the participants at baseline to fill out a short questionnaire regarding (1) receiving information; (2) motivation, perseverance, and discipline; and (3) acceptance that unbeneficial but often enjoyable habit patterns need to change. Participants rank their motivation, perseverance, and discipline on a visual analogue scale (VAS) from 0 to 10 (with 0 representing no/lowest, 5 moderate, and 10 highest motivation, perseverance, and/or discipline). Subsequently, individual lifestyle goal(s) on one (or more) of the four lifestyle components (smoking, food pattern, weight, and/or physical activity) are defined. The following steps determine ambivalence/assessment status: the stages of change model (the transtheoretical Model of Change) [39,41]; and the 'DARN-CAT' motivational interviewing model (increasing/recognizing of 'change talk' using the six signs in the individual's language) [40,41]. The contents of the coaching depend on the stage and/or need of the participant and lifestyle assessments. Documents with assignments and tables (e.g., specific foods and levels/concentrations of beneficial AMD-nutrients) are provided to enhance awareness and stimulate habitual change during these individual coaching sessions.

To measure, score, and visualize progression or relapse, and to enhance adherence during the coaching sessions, we developed a 13-point easy-to-use lifestyle scoring form (Figure S1) and a visualization form (Figure S2) for follow up. Both forms include the total GRS (Figure S1 page 2) and lifestyle information. Lifestyle risk estimates are based on validated studies (i.e., smoking, BMI, MeDiet (13 items), physical activity). Each healthy outcome receives a point. The total score can range from 0 to 13, in which high scores represent an AMD-healthy lifestyle (Figures S1 and S2).

The lifestyle score form is used at baseline for group B and C, and at each additional time point (mos. 0, 3, 6, 9, 12, and 24) for Group C. At the second (mo. 12) and third (mo. 24) visit at our research center, all participants receive the complete lifestyle score form and visualization thereof at mos. 0, 12, and 24, respectively. For a more detailed description, see the Supplementary Materials.

3. Study Outcomes

The study outcomes are as follows: adherence to lifestyle recommendations; identify change over time in biomarkers from blood related to AMD lifestyle determinants; the feasibility and acceptability of participants compared across the three intervention strategies, as based on quantitative and qualitative data.

3.1. *Primary Study Outcome*

Improvement of the overall lifestyle score was measured by the 13-point form (see previous paragraph and Figure S1: Lifestyle scoring form), which is based on the questionnaires inquiring about smoking habits, BMI, MeDiet, and physical activity taken at baseline and 12 months (see previous paragraph data collection and measurement).

3.2. *Secondary Study Endpoints of AMD-Life*

Improvement of the overall lifestyle score at 24 months.

Change of levels of blood biomarkers from serum/plasma and finger-prick at 12 and 24 months.

Personal feasibility and acceptability of lifestyle change measured by questionnaires at 3, 6, 9, 12, and 24 months.

Changes in macular pigment optical density as measured with MPOD at 12 and 24 months.

Changes in quantification of features (e.g., drusen) and stage of AMD as measured on multimodal images at 12 and 24 months.

Other Study Parameters

Other study parameters are as follows: age, gender, blood pressure, hypertension, education and socioeconomic factors, comorbidity such as cardiovascular disease, medication, genetic susceptibility of AMD evaluated by GRS, major risk and protective variants, and rare variants.

3.3. *Statistical Analyses*

To compare baseline characteristics between treatment groups we will use the Chi square test and ANOVA. We will conduct longitudinal analysis on the primary and secondary outcomes, with treatment groups as independent variables. To account for clustering within the individual due to repeated measurements, we will use mixed-effects models. We will conduct 'intention to treat' analysis and per protocol as a sensitivity analysis.

All statistical analyses will be carried out using R, version 3.5.3 (R Core Team, 2016) and SPSS Statistics, version 29.0.0.0 (IBM, Armonk, NY, USA).

4. Discussion

Achieving sustainable lifestyle change in AMD patients is challenging in daily practice. The goal of the AMD-Life study is to define the best strategy to accomplish awareness and sustainable compliance to recommendations by participants diagnosed with AMD. The study is a 1 year, three-arm (group A, B, and C), self-contained open-label randomized clinical trial, followed by an observation period in year 2.

An important strength of our study is the very detailed exploratory investigation, such as the block-wise randomized design, the close monitoring of participants, and the extensive questionnaires about the major lifestyle exposures for AMD, smoking, dietary pattern, BMI, and physical activity. We use the validated 389-dietary FFQ to measure dietary behavior and to record frequency of each intake as times per day, per week, or per month. Thereby, we are able to calculate different dietary pattern scoring methods, such as the Mediterranean diet score to achieve insight on adherence to food categories, and we are able to calculate single nutrient intake. The 16-point MeDiet screening instrument allows quick analysis of food intake and prompt determination of recommendations for change. Smoking is a well-established risk factor, but an infrequent exposure in this age-category; BMI and physical activity are less well established for AMD, but obesity and sedentary behavior are much more prevalent. An approach that combines these four exposures is likely to be more successful than a focus on only one exposure. Other strengths include extensive phenotyping of AMD features and a large set of other covariates for analysis, including biomarkers. The biomarker panel that will be assessed is very extensive, with the goal of exploring associations related to lifestyle and AMD. It involves complement factors in the active and nonactive state, oxidative stress markers, and markers of inflammation, lipids, anti-oxidants, fat soluble vitamins, minerals, and fatty acids. When confirmed, these biomarkers can serve as guidance and motivation for lifestyle change. The patient will be 'in control' of his/her own disease and have a better understanding of the consequences for the risks of his/her behavior.

Potential limitations of this study are the relatively small number of study participants. This study is designed as a pilot (inception phase) study. During this phase, we will set up the main elements, test our assumptions and the feasibility, check procedures, focus on adherence, and report participant and physician experience for three different strategies for stimulating a healthy lifestyle in AMD patients, specifically targeted at long-term behavioral change. Another limitation is the use of self-reported physical activity. More objective measurements could be obtained using triaxial accelerometers. A disadvantage of block randomization is that the allocation of participants could result in selection bias [56]. To reduce this bias, investigators are blinded to the randomization procedure.

Meta-analyses show that motivational interviewing (MI) is effective for decreasing alcohol and drug use in adults and adolescents, and evidence is accumulating in other areas of health, including smoking cessation, improving adherence to treatment and medication, and diabetes management [57–60]. The AMD-Life study will assess if coaching based on the behavioral change technique (BCT) [38,39] and MI [40,41] can be of additional value to change habitual behaviors and adhere to the AMD recommendations. Because implementation of a lifestyle coaching program in AMD management will be challenging, we evaluate changes in habitual behavior and assess adherence at different time points (mos. 3, 6, 9, and 12); this will provide useful information for the implementation of lifestyle coaching programs.

Over 200 clinical trials are taking place targeting various processes in the pathogenic pathway (e.g., complement activation). Aside from the direct costs, these trials also take a toll on patients, which need to be enrolled in large numbers and adhere to these trials without a warranty of benefit. It is not a secret that 90% of drug candidates in clinical trials fail, and patients who have entered trials before are usually excluded from new ones. If trials manage to be successful, new drugs rarely have the potential to reduce the risk of disease progression by half. Lifestyle intervention has this potential. AMD-Life, when validated, will provide the evidence that lifestyle changes can be achieved in persons at risk of AMD progression.

The psychological and social impact of AMD on quality of life is significant, and health-care costs related to vision loss are expected to increase exponentially over the following years [61]. Annual costs of care and treatment of AMD in Europe currently exceed €75 billion; macroeconomic costs related to visual disability can be up to €100 billion. The AMD-Life study fits well within the changing paradigm of patient and healthcare management. In general, clinical focus is moving from medical treatment of end-stage diseases to preventative actions at early stages. Insight into which strategy (A, B, or C) will stimulate preventative actions and adherence to clinical recommendations will not only slow down disease progression but also support wider clinical application of lifestyle intervention and subsequently lower patient burden and health care costs.

5. Conclusions

Patients at risk for AMD show low general adherence to healthy diets or regular intake of supplements. ADM-Life was set up to research and assess in clinical application whether adherence improves if patients experience a direct reward or receive feedback of the benefits of their behavior change to general health. The ultimate goal is to reduce the risk of late AMD and thereby improve individual lives and reduce blindness and huge health care costs.

Supplementary Materials: The following supporting information can be downloaded at: https://www.mdpi.com/article/10.3390/nu15030602/s1, † Summary of the AMD-Life team members; Tabel S1: Oral supplements (AREDS 2 formula & Omega 3 forte fish oil) provided to participants in AMD-Life study for 1 year; Tabel S2: Single Nucleotide Polymorphisms used for AMD-Life GRS calculation; Figure S1: Three page, easy-to-use scoring form used in the AMD-Life study to inform participants on their lifestyle and genetic risk and to stimulate lifestyle change; Figure S2: Visualization of the lifestyle scoring over time. The overview graph visualizes the current status of the lifestyle and enables to display follow up measurement.

Author Contributions: Conceptualization, A.P.M.d.K.-B., C.C.W.K. and M.A.M.-S.; Methodology, A.P.M.d.K.-B., C.C.W.K., M.A.M.-S., J.C.K.-d.J., A.G.U., J.G.J.v.R. and members of AMD-Life team, analysis, A.P.M.d.K.-B., J.C.K.-d.J., J.G.J.v.R. and members of the AMD-Life team, writing—original draft preparation, A.P.M.d.K.-B.; writing—review and editing, A.P.M.d.K.-B., C.C.W.K., M.A.M.-S., J.C.K.-d.J., T.G.V., A.G.U. and members of the AMD-Life team; supervision and funding, C.C.W.K. and M.A.M.-S. All authors have read and agreed to the published version of the manuscript.

Funding: This research was supported by Combined Ophthalmic Research Rotterdam (CORR), Rotterdamse Stichting Blindenbelangen (RSB), Vitaminen op Recept, and by the following foundations that contributed through Uitzicht: Oogfonds, Vereniging Bartiméus Sonneheerdt, Algemene Nederlandse Vereniging ter Voorkoming van Blindheid, and Landelijke Stichting voor Blinden en Slechtzienden (Uitzicht 2016–19). The funding organizations had no role in the design or conduct of this research.

Institutional Review Board Statement: Ethical approval was obtained from the Medical Ethical Committee at Erasmus MC (MEC-2018-063 and amendment NL65052.078.18 (27 July 2021)).

Informed Consent Statement: Prior to the study, informed consent will be obtained from all participant involved in the study.

Data Availability Statement: Data are held in at the Erasmus Medical Center and are available upon written request.

Acknowledgments: The authors thank all volunteers for their participation in the AMD-Life Study; the staff of the AMD-Life Study; all physician at outpatient clinics, the Dutch Macular Association i.e., for referring potential participants; Peter C. de Koning; Heleen Schoots; Bram Harder; Karin Gordijn; Emar Vogelaar; Erwin vd Horst; Bernard Kuijer; Thomas E. Gundersen; Siv Kaland; Jaco Huijzers; Ian J. Murray; Roel Faber; Selda Eskikoy–Cazoglu; Lorissa Holkema.

Conflicts of Interest: The authors declare no conflict of interest.

References

1. Wong, W.L.; Su, X.; Li, X.; Cheung, C.M.G.; Klein, R.; Cheng, C.-Y.; Wong, T.Y. Global prevalence of age-related macular degeneration and disease burden projection for 2020 and 2040: A systematic review and meta-analysis. *Lancet Glob. Health* **2014**, *2*, e106–e116. [CrossRef]
2. Owen, C.G.; Jarrar, Z.; Wormald, R.; Cook, D.G.; Fletcher, A.E.; Rudnicka, A.R. The estimated prevalence and incidence of late stage age related macular degeneration in the UK. *Br. J. Ophthalmol.* **2012**, *96*, 752–756. [CrossRef]
3. Colijn, J.M.; Buitendijk, G.H.S.; Prokofyeva, E.; Alves, D.; Cachulo, M.L.; Khawaja, A.P.; Cougnard-Gregoire, A.; Merle, B.M.J.; Korb, C.; Erke, M.G.; et al. Prevalence of Age-Related Macular Degeneration in Europe: The Past and the Future. *Ophthalmology* **2017**, *124*, 1753–1763. [CrossRef] [PubMed]
4. van Leeuwen, R.; Klaver, C.C.; Vingerling, J.R.; Hofman, A.; de Jong, P.T. The risk and natural course of age-related maculopathy: Follow-up at 6 1/2 years in the Rotterdam study. *Arch. Ophthalmol.* **2003**, *121*, 519–526. [CrossRef] [PubMed]
5. Mehta, H.; Tufail, A.; Daien, V.; Lee, A.Y.; Nguyen, V.; Ozturk, M.; Barthelmes, D.; Gillies, M.C. Real-world outcomes in patients with neovascular age-related macular degeneration treated with intravitreal vascular endothelial growth factor inhibitors. *Prog. Retin. Eye Res.* **2018**, *65*, 127–146. [CrossRef]
6. Liao, D.S.; Grossi, F.V.; El Mehdi, D.; Gerber, M.R.; Brown, D.M.; Heier, J.S.; Wykoff, C.C.; Singerman, L.J.; Abraham, P.; Grassmann, F.; et al. Complement C3 Inhibitor Pegcetacoplan for Geographic Atrophy Secondary to Age-Related Macular Degeneration: A Randomized Phase 2 Trial. *Ophthalmology* **2020**, *127*, 186–195. [CrossRef]
7. Fritsche, L.G.; Igl, W.; Bailey, J.N.; Grassmann, F.; Sengupta, S.; Bragg-Gresham, J.L.; Burdon, K.P.; Hebbring, S.J.; Wen, C.; Gorski, M.; et al. A large genome-wide association study of age-related macular degeneration highlights contributions of rare and common variants. *Nat. Genet.* **2016**, *48*, 134–143. [CrossRef]
8. Colijn, J.M.; Meester-Smoor, M.; Verzijden, T.; de Breuk, A.; Silva, R.; Merle, B.M.J.; Cougnard-Grégoire, A.; Hoyng, C.B.; Fauser, S.; Coolen, A.; et al. Genetic Risk, Lifestyle, and Age-Related Macular Degeneration in Europe: The EYE-RISK Consortium. *Ophthalmology* **2021**, *128*, 1039–1049. [CrossRef]
9. Geerlings, M.J.; de Jong, E.K.; den Hollander, A.I. The complement system in age-related macular degeneration: A review of rare genetic variants and implications for personalized treatment. *Mol. Immunol.* **2017**, *84*, 65–76. [CrossRef]
10. Corominas, J.; Colijn, J.M.; Geerlings, M.J.; Pauper, M.; Bakker, B.; Amin, N.; Lores Motta, L.; Kersten, E.; Garanto, A.; Verlouw, J.A.M.; et al. Whole-Exome Sequencing in Age-Related Macular Degeneration Identifies Rare Variants in COL8A1, a Component of Bruch's Membrane. *Ophthalmology* **2018**, *125*, 1433–1443. [CrossRef]
11. Huan, T.; Cheng, S.Y.; Tian, B.; Punzo, C.; Lin, H.; Daly, M.; Seddon, J.M. Identifying Novel Genes and Variants in Immune and Coagulation Pathways Associated with Macular Degeneration. *Ophthalmol. Sci.* **2023**, *3*, 100206. [CrossRef]
12. Vingerling, J.R.; Hofman, A.; Grobbee, D.E.; de Jong, P.T. Age-related macular degeneration and smoking. The Rotterdam Study. *Arch. Ophthalmol.* **1996**, *114*, 1193–1196. [CrossRef] [PubMed]
13. Chakravarthy, U.; Augood, C.; Bentham, G.C.; de Jong, P.T.; Rahu, M.; Seland, J.; Soubrane, G.; Tomazzoli, L.; Topouzis, F.; Vingerling, J.R.; et al. Cigarette smoking and age-related macular degeneration in the EUREYE Study. *Ophthalmology* **2007**, *114*, 1157–1163. [CrossRef] [PubMed]
14. Weeks, D.E.; Conley, Y.P.; Tsai, H.J.; Mah, T.S.; Schmidt, S.; Postel, E.A.; Agarwal, A.; Haines, J.L.; Pericak-Vance, M.A.; Rosenfeld, P.J.; et al. Age-related maculopathy: A genomewide scan with continued evidence of susceptibility loci within the 1q31, 10q26, and 17q25 regions. *Am. J. Hum. Genet.* **2004**, *75*, 174–189. [CrossRef]
15. Seddon, J.M.; Francis, P.J.; George, S.; Schultz, D.W.; Rosner, B.; Klein, M.L. Association of CFH Y402H and LOC387715 A69S with progression of age-related macular degeneration. *Jama* **2007**, *297*, 1793–1800. [CrossRef] [PubMed]
16. de Koning-Backus, A.P.M.; Buitendijk, G.H.S.; Kiefte-de Jong, J.C.; Colijn, J.M.; Hofman, A.; Vingerling, J.R.; Haverkort, E.B.; Franco, O.H.; Klaver, C.C.W. Intake of Vegetables, Fruit, and Fish is Beneficial for Age-Related Macular Degeneration. *Am. J. Ophthalmol.* **2019**, *198*, 70–79. [CrossRef] [PubMed]
17. Merle, B.M.; Silver, R.E.; Rosner, B.; Seddon, J.M. Adherence to a Mediterranean diet, genetic susceptibility, and progression to advanced macular degeneration: A prospective cohort study. *Am. J. Clin. Nutr.* **2015**, *102*, 1196–1206. [CrossRef] [PubMed]
18. Merle, B.M.J.; Colijn, J.M.; Cougnard-Grégoire, A.; de Koning-Backus, A.P.M.; Delyfer, M.N.; Kiefte-de Jong, J.C.; Meester-Smoor, M.; Féart, C.; Verzijden, T.; Samieri, C.; et al. Mediterranean Diet and Incidence of Advanced Age-Related Macular Degeneration: The EYE-RISK Consortium. *Ophthalmology* **2019**, *126*, 381–390. [CrossRef]
19. McGuinness, M.B.; Le, J.; Mitchell, P.; Gopinath, B.; Cerin, E.; Saksens, N.T.M.; Schick, T.; Hoyng, C.B.; Guymer, R.H.; Finger, R.P. Physical Activity and Age-related Macular Degeneration: A Systematic Literature Review and Meta-analysis. *Am. J. Ophthalmol.* **2017**, *180*, 29–38. [CrossRef] [PubMed]
20. Merle, B.M.J.; Buaud, B.; Korobelnik, J.F.; Bron, A.; Delyfer, M.N.; Rougier, M.B.; Savel, H.; Vaysse, C.; Creuzot-Garcher, C.; Delcourt, C. Plasma long-chain omega-3 polyunsaturated fatty acids and macular pigment in subjects with family history of age-related macular degeneration: The Limpia Study. *Acta Ophthalmol.* **2017**, *95*, e763–e769. [CrossRef] [PubMed]
21. Souied, E.H.; Delcourt, C.; Querques, G.; Bassols, A.; Merle, B.; Zourdani, A.; Smith, T.; Benlian, P. Oral docosahexaenoic acid in the prevention of exudative age-related macular degeneration: The Nutritional AMD Treatment 2 study. *Ophthalmology* **2013**, *120*, 1619–1631. [CrossRef]
22. FAO. Food and Agriculture Organization of the United Nations: Food-Based Dietary Guidelines. Available online: http://www.fao.org/nutrition/education/food-based-dietary-guidelines/regions/countries (accessed on 1 October 2022).

23. Trichopoulou, A.; Kouris-Blazos, A.; Wahlqvist, M.L.; Gnardellis, C.; Lagiou, P.; Polychronopoulos, E.; Vassilakou, T.; Lipworth, L.; Trichopoulos, D. Diet and overall survival in elderly people. *BMJ (Clin. Res. Ed.)* **1995**, *311*, 1457–1460. [CrossRef]
24. Trichopoulou, A.; Costacou, T.; Bamia, C.; Trichopoulos, D. Adherence to a Mediterranean diet and survival in a Greek population. *New Engl. J. Med.* **2003**, *348*, 2599–2608. [CrossRef] [PubMed]
25. Organisation, W.H. Healthy Diet. Available online: https://www.who.int/news-room/fact-sheets/detail/healthy-diet (accessed on 1 October 2022).
26. CBS and RIVM. Gezondheidsenquette/Leefstijlmonitor. Available online: https://www.cbs.nl/nl-nl/nieuws/2015/17/nederland-eet-onvoldoende-groente-fruit-en-vis (accessed on 1 January 2016).
27. Laurenti, G. *1961–2003 Fish and Fishery Products: World Apparent Consumption Statistics Based on Food Balance Sheets*; FAO Fisheries Circular. No. 821, Rev. 8; FAO: Rome, Italy, 2007; p. 429.
28. Age-Related Eye Disease Study Research Group. The Age-Related Eye Disease Study (AREDS): Design implications. AREDS report no. 1. *Control. Clin. Trials* **1999**, *20*, 573–600. [CrossRef] [PubMed]
29. Chew, E.Y.; SanGiovanni, J.P.; Ferris, F.L.; Wong, W.T.; Agron, E.; Clemons, T.E.; Sperduto, R.; Danis, R.; Chandra, S.R.; Blodi, B.A.; et al. Lutein/zeaxanthin for the treatment of age-related cataract: AREDS2 randomized trial report no. 4. *JAMA Ophthalmol.* **2013**, *131*, 843–850. [CrossRef] [PubMed]
30. Lin, P.; McClintic, S.M.; Nadeem, U.; Skondra, D. A Review of the Role of the Intestinal Microbiota in Age-Related Macular Degeneration. *J. Clin. Med.* **2021**, *10*, 2072. [CrossRef]
31. Hochstetler, B.S.; Scott, I.U.; Kunselman, A.R.; Thompson, K.; Zerfoss, E. Adherence to recommendations of the age-related eye disease study in patients with age-related macular degeneration. *Retina* **2010**, *30*, 1166–1170. [CrossRef]
32. Ng, W.T.; Goggin, M. Awareness of and compliance with recommended dietary supplement among age-related macular degeneration patients. *Clin. Exp. Ophthalmol.* **2006**, *34*, 9–14. [CrossRef] [PubMed]
33. Shah, S.U.; Pilli, S.; Telander, D.G.; Morse, L.S.; Park, S.S. Reprint of: Survey of patients with age-related macular degeneration: Knowledge and adherence to recommendations. *Can. J. Ophthalmol.* **2015**, *50* (Suppl. S1), S23–S28. [CrossRef]
34. Sahu, A.; Edwards, R.; Harrison, R.A.; Thornton, J.; Kelly, S.P. Attitudes and behaviour of ophthalmologists to smoking cessation. *Eye* **2008**, *22*, 246–250. [CrossRef]
35. Lawrenson, J.G.; Evans, J.R. Advice about diet and smoking for people with or at risk of age-related macular degeneration: A cross-sectional survey of eye care professionals in the UK. *BMC Public Health* **2013**, *13*, 564. [CrossRef] [PubMed]
36. Cohen, D.J.; Clark, E.C.; Lawson, P.J.; Casucci, B.A.; Flocke, S.A. Identifying teachable moments for health behavior counseling in primary care. *Patient Educ. Couns.* **2011**, *85*, e8–e15. [CrossRef]
37. Brust, M.; Gebhardt, W.A.; Numans, M.E.; Kiefte-de Jong, J.C. Teachable moments: The right moment to make patients change their lifestyle. *Ned. Tijdschr. Geneeskd.* **2020**, *164*, D4835. [PubMed]
38. Kwasnicka, D.; Dombrowski, S.U.; White, M.; Sniehotta, F. Theoretical explanations for maintenance of behaviour change: A systematic review of behaviour theories. *Health Psychol. Rev.* **2016**, *10*, 277–296. [CrossRef] [PubMed]
39. Prochaska, J.O.; Diclemente, C.C. Toward a Comprehensive Model of Change. In *Treating Addictive Behaviors: Processes of Change*; Miller, W.R., Heather, N., Eds.; Springer: Boston, MA, USA, 1986; pp. 3–27.
40. Miller, W.R.; Rollnick, S. *Motivational Interviewing, Third Edition: Helping People Change*, 3rd ed.; Guilford Press: New York, NY, USA, 2012.
41. DiClemente, C.; Velasquez, M. Motivational interviewing and the stages of change. In *Motivational Interviewing: Preparing People for Change*; Miller, W., Rollnick, S., Eds.; Guilford Press: New York, NY, USA, 2002; pp. 201–216.
42. Klein, R.; Davis, M.D.; Magli, Y.L.; Segal, P.; Klein, B.E.; Hubbard, L. The Wisconsin age-related maculopathy grading system. *Ophthalmology* **1991**, *98*, 1128–1134. [CrossRef]
43. EyeNED Reading Center. Available online: https://www.eyened.nl/ (accessed on 1 January 2023).
44. Gattoussi, S.; Buitendijk, G.H.S.; Peto, T.; Leung, I.; Schmitz-Valckenberg, S.; Oishi, A.; Wolf, S.; Deák, G.; Delcourt, C.; Klaver, C.C.W.; et al. The European Eye Epidemiology spectral-domain optical coherence tomography classification of macular diseases for epidemiological studies. *Acta Ophthalmol.* **2019**, *97*, 364–371. [CrossRef] [PubMed]
45. Staurenghi, G.; Sadda, S.; Chakravarthy, U.; Spaide, R.F. Proposed lexicon for anatomic landmarks in normal posterior segment spectral-domain optical coherence tomography: The IN•OCT consensus. *Ophthalmology* **2014**, *121*, 1572–1578. [CrossRef] [PubMed]
46. Fallaize, R.; Livingstone, K.M.; Celis-Morales, C.; Macready, A.L.; San-Cristobal, R.; Navas-Carretero, S.; Marsaux, C.F.M.; O'Donovan, C.B.; Kolossa, S.; Moschonis, G.; et al. Association between Diet-Quality Scores, Adiposity, Total Cholesterol and Markers of Nutritional Status in European Adults: Findings from the Food4Me Study. *Nutrients* **2018**, *10*, 49. [CrossRef]
47. Sakhi, A.K.; Bastani, N.E.; Ellingjord-Dale, M.; Gundersen, T.E.; Blomhoff, R.; Ursin, G. Feasibility of self-sampled dried blood spot and saliva samples sent by mail in a population-based study. *BMC Cancer* **2015**, *15*, 265. [CrossRef]
48. Goldbohm, R.A.; van den Brandt, P.A.; Brants, H.A.; van't Veer, P.; Al, M.; Sturmans, F.; Hermus, R.J. Validation of a dietary questionnaire used in a large-scale prospective cohort study on diet and cancer. *Eur. J. Clin. Nutr.* **1994**, *48*, 253–265. [PubMed]
49. van Lee, L.; Geelen, A.; van Huysduynen, E.J.; de Vries, J.H.; van't Veer, P.; Feskens, E.J. The Dutch Healthy Diet index (DHD-index): An instrument to measure adherence to the Dutch Guidelines for a Healthy Diet. *Nutr. J.* **2012**, *11*, 49. [CrossRef] [PubMed]

50. NEVO-Tabel; Dutch Food Composition Database 2011, R.V., Den Haag. 2011. Available online: https://nevo-online.rivm.nl/Home/En (accessed on 1 October 2022).
51. Klipstein-Grobusch, K.; den Breeijen, J.H.; Goldbohm, R.A.; Geleijnse, J.M.; Hofman, A.; Grobbee, D.E.; Witteman, J.C. Dietary assessment in the elderly: Validation of a semiquantitative food frequency questionnaire. *Eur. J. Clin. Nutr.* **1998**, *52*, 588–596. [CrossRef] [PubMed]
52. Feunekes, G.I.; Van Staveren, W.A.; De Vries, J.H.; Burema, J.; Hautvast, J.G. Relative and biomarker-based validity of a food-frequency questionnaire estimating intake of fats and cholesterol. *Am. J. Clin. Nutr.* **1993**, *58*, 489–496. [CrossRef]
53. Martínez-González, M.A.; Fernández-Jarne, E.; Serrano-Martínez, M.; Wright, M.; Gomez-Gracia, E. Development of a short dietary intake questionnaire for the quantitative estimation of adherence to a cardioprotective Mediterranean diet. *Eur. J. Clin. Nutr.* **2004**, *58*, 1550–1552. [CrossRef] [PubMed]
54. Schröder, H.; Fitó, M.; Estruch, R.; Martínez-González, M.A.; Corella, D.; Salas-Salvadó, J.; Lamuela-Raventós, R.; Ros, E.; Salaverría, I.; Fiol, M.; et al. A short screener is valid for assessing Mediterranean diet adherence among older Spanish men and women. *J. Nutr.* **2011**, *141*, 1140–1145. [CrossRef] [PubMed]
55. Meyer, A.M.; Evenson, K.R.; Morimoto, L.; Siscovick, D.; White, E. Test-retest reliability of the Women's Health Initiative physical activity questionnaire. *Med. Sci. Sports Exerc.* **2009**, *41*, 530–538. [CrossRef] [PubMed]
56. Efird, J. Blocked randomization with randomly selected block sizes. *Int. J. Environ. Res. Public Health* **2011**, *8*, 15–20. [CrossRef]
57. Samdal, G.B.; Eide, G.E.; Barth, T.; Williams, G.; Meland, E. Effective behaviour change techniques for physical activity and healthy eating in overweight and obese adults; systematic review and meta-regression analyses. *Int. J. Behav. Nutr. Phys. Act.* **2017**, *14*, 42. [CrossRef]
58. Lundahl, B.W.; Kunz, C.; Brownell, C.; Tollefson, D.; Burke, B.L. A Meta-Analysis of Motivational Interviewing: Twenty-Five Years of Empirical Studies. *Res. Soc. Work. Pract.* **2010**, *20*, 137–160. [CrossRef]
59. Gray, E.; McCambridge, J.; Strang, J. The effectiveness of motivational interviewing delivered by youth workers in reducing drinking, cigarette and cannabis smoking among young people: Quasi-experimental pilot study. *Alcohol Alcohol.* **2005**, *40*, 535–539. [CrossRef]
60. Fortier, M.S.; Duda, J.L.; Guerin, E.; Teixeira, P.J. Promoting physical activity: Development and testing of self-determination theory-based interventions. *Int. J. Behav. Nutr. Phys. Act.* **2012**, *9*, 20. [CrossRef] [PubMed]
61. Frost; Sullivan. The Economic Benefits of Using Lutein and Zeaxanthin Food Supplements in the European Union. Available online: https://www.frost.com/frost-perspectives/the-economic-benefits-of-using-lutein-and-zeaxanthin-food-supplements-in-the-european-union/ (accessed on 1 October 2022).

nutrients

MDPI

Article

Relationship of the Behavior of Older Participants with Body Composition Change: Results of the Second Wave of the Cognition of Older People, Education, Recreational Activities, Nutrition, Comorbidities, and Functional Capacity Studies (COPERNICUS)

Agnieszka Kujawska [1], Guillermo F. López Sánchez [2], Flaka Hoti [3], Sławomir Kujawski [1,*], Paweł Zalewski [1,4] and Kornelia Kędziora-Kornatowska [5]

1 Department of Exercise Physiology and Functional Anatomy, Ludwik Rydygier Collegium Medicum in Bydgoszcz, Nicolaus Copernicus University in Toruń, Świętojańska 20, 85-077 Bydgoszcz, Poland; agnieszka.kujawska@cm.umk.pl (A.K.)
2 Division of Preventive Medicine and Public Health, Department of Public Health Sciences, School of Medicine, University of Murcia, 30120 Murcia, Spain
3 Department of Health Management, University of Gjakova "Fehmi Agani", 50000 Gjakova, Kosovo
4 Laboratory of Centre for Preclinical Research, Department of Experimental and Clinical Physiology, Medical University of Warsaw, 1b Banacha Street, 02-097 Warsaw, Poland
5 Department of Geriatrics, Collegium Medicum in Bydgoszcz, Nicolaus Copernicus University in Toruń, 85-094 Bydgoszcz, Poland
* Correspondence: skujawski@cm.umk.pl

Abstract: Background: To examine the relationship between the frequency of physical activities and food product consumption with body composition change after two years in a sample of older people. Methods: Body composition, mass change, frequency of physical activity, and food products consumption were measured. Depression severity, health self-assessment, cognitive function, and demographic data were included as confounders. Results: There were no significant changes in body composition except for a reduction in visceral fat level within two years ($p < 0.05$). Drinking beer and eating sweets a few times per week were associated with a significant increase in body fat percentage ($p < 0.05$). Drinking green or white tea more frequently than a few times per year was related to an increase in body fat (3.18 to 3.88%, $p < 0.05$). Contrarily, daily consumption of coffee was related to a decrease in body fat ($p = 0.029$). Subjects who ate sweets once a week or more frequently consumed coffee more often. Conclusions: More frequent drinking of beer or of green or white tea and consumption of sweets were related to an increase in body fat percentage, while daily coffee consumption was related to a decrease in body fat percentage after two years in older, healthy subjects. Noteworthily, the frequencies of food product consumption are interrelated.

Keywords: diet; aging; exercise; geroprotectors; healthspan; network analysis

1. Introduction

Aging is the main risk factor for most chronic disorders and conditions that decrease health span [1]. Over the past 40 years, obesity has become more common, especially among individuals 60 to 74 years old [2]. Biological mechanisms related to aging are highly complex [3]. Furthermore, it seems that aging and multiple chronic disorders share common mechanisms [1]. These might include (mal)adaptation to stress, loss of proteostasis, stem cell exhaustion, metabolism derangement, macromolecular damage, aberrant epigenetic modifications, and inflammation [4]. Obesity is linked to impairment and the aggravation of chronic conditions such as type 2 diabetes, cardiovascular disease, and osteoarthritis in older persons [5]. The inflammatory environment that is present in obesity and metabolic syndrome is present in age-related disorders including sarcopenia, frailty, and dementia as well; as a result, these two sets of pathologies feed off of one another [6–9].

In addition to the interrelationship of pathologies with each other, there are frequent interrelationships between two or more health behaviors of individuals that have a positive or negative effect on health depending on the type of behavior. For example, a study conducted in Japan by Yamauchi et al. (2007) [10], which included older individuals and sought to examine the relationship between physical activity and healthy dietary habits, reported a positive correlation between these two health behaviors. Thus, many studies report positive combined practices (i.e., a link between positive health behaviors), as shown in a study by Hu et al. [11], in which a high level of physical activity, adequate diet, and rest were reported in combination by a large proportion of participants. The reasons why the practice of one health behavior may influence the practice of another health behavior are not entirely clear and remain to be explored; however, self-efficacy theory is often cited as an important factor in this matter. When a person succeeds in changing one lifestyle behavior, he or she becomes more confident in practicing other healthy behaviors [12].

The network approach is an important approach to deeper identification and explanation of the various relationships. Numerous studies can be found in the literature dealing with the application of the network approach in different fields of medical sciences [13,14]. According to Havey [15], a network refers to various structures consisting of variables represented by nodes and the relationships (formally called edges) between nodes. A network refers to different structures made up of more or less interconnected variables. Therefore, networks allow us to have a deeper understanding of how interconnected variables are placed and affect an individual's health state. Networks evolve and respond to the changes that the variables undergo, and these complex systems of connectivity between variables can only be revealed by network analysis [15].

This study aims to examine the relationship between the frequency of physical activities and food product consumption with body composition (body fat percentage, skeletal muscle, mass, and visceral fat) changes within two years in a sample of older people.

2. Materials and Methods

2.1. Study Group

A regional TV and radio marketing campaign, health-related lectures at Collegium Medicum University, several older people's organizations in Bydgoszcz, day care facilities for the elderly, and various senior meeting groups were used to recruit participants for this study. Information on the research included a description of an opportunity to undergo free of charge physical, physiotherapeutic, nutritional, social, and cognitive examination for participants 60 years of age and older. Participants were recruited for the current study between November 2015 and February 2018.

The only exclusion criterion from the study was age under 60. The examination was conducted in the Department of Geriatrics, Collegium Medicum University Hospital in Bydgoszcz, Poland. The study was approved by the Ethics Committee, Ludwik Rydygier Memorial Collegium Medicum in Bydgoszcz, Nicolaus Copernicus University, Torun (KB 340/2015). Written, informed consent was obtained from all participants.

2.2. Assessment Methods

2.2.1. Body Composition Analysis

Analyzing body composition and body mass was performed using the Tanita BC-545 body fat analyzer. When being weighed, participants wore light clothing. Accuracy in weighing was 0.1%. Bioelectric impedance analysis (BIA) was used to quantify the body composition indices of body fat (%), visceral fat (units), and muscle mass (kg). All parameters were measured using built-in algorithms. Whole-body fat and lean mass measurement using Tanita BC-545 have a good agreement in comparison to dual energy X-Ray absorptiometry [16]. The error of measurement using BIA might be related to nutrition status, tissue temperature, and hydration [17]. As examinations lasted for about three hours per patient, subjects were reminded to take food and drink with them and consume them if needed. In addition, an opportunity to use the toilet with a reminder was provided. Participants were instructed to stand with electrodes on their feet and their hands. Respondents provided information on their height to shorten the time spent on examination; body mass index (BMI) was calculated using WHO norms [18]. Normal body type was chosen for all participants.

2.2.2. Activity Level Questionnaire

Utilizing a previously detailed questionnaire, the frequency of current physical, mental, and social activities was evaluated [19]. Questions involved the following physical activities: short walks around the house, long walks, gymnastics, cycling, running/jogging, swimming, skiing, team games, sailing, horse riding, Nordic walking, tennis/table tennis, dancing, and work on a garden plot or mushroom collection. The following questions on the frequency of food product consumption were asked: drinking beer, green/white tea, red wine, vodka, and other 40% (80-proof) alcohols, and eating chocolate, sweets, fish, coffee, vegetables, and fruit. Frequency was coded into seven categories ("never", "once a year", "several times a year", "1–2 times a month", "once a week", "a few times a week", and "daily").

2.2.3. Depression Severity and Health Assessment

The Geriatric Depression Scale (GDS), composed of 15 items, was utilized as a depression screening tool [20]. It has been demonstrated that this form can be useful in very old persons with and without cognitive impairment [21]. Questions concerned the quality of life (e.g., do you feel full of energy?), current circumstances (e.g., do you feel that your situation is hopeless?), daily activities (e.g., have you dropped many of your activities and interests?), mental health (e.g., do you feel happy most of the time?), and outlook on life (e.g., are you basically satisfied with your life?). A GDS-15 score ranges from 0 to 15 points, with a higher score reflecting greater severity of depressive symptoms

In addition, subjects were asked to carry out a self-assessment of their current health state on a 10-point scale, with 10 meaning a perfect state of health.

2.2.4. Cognitive Function Assessment

The Montreal Cognitive Assessment (MoCA) was used to measure general cognitive function [22]. The MoCA evaluates each major cognitive function domains, including executive function, short-term memory recall, and visuospatial abilities. The former domain was examined by a mini-form of TMT B, a phonemic fluency task, and a two-item verbal abstraction task. MoCA scores range from 0 to 30 points, with a higher score denoting better global cognitive function.

2.2.5. Demographic and Occupation-Related Data

A number of years spent on education was measured using a self-report questionnaire. Occupational status was at first categorized as follows: white-collar worker, a white-collar worker in a managerial position, owner of a craft/entrepreneurial enterprise, military/policeman/other uniformed service, seller/employee of trade, farmer in an individual farm, physical worker (qualified), and unskilled worker. Then, the last three categories were unified as "low occupational status" and the rest as "high occupational status". Eventually, a dichotomous variable with the highest occupational status achieved during the course of the respondent's career was created [19].

2.3. *Statistical Analysis*

Dependent t-tests were used to analyse whether assumptions were met; otherwise, Wilcoxon tests were used to compare results from repeated measurements (before vs. after two years). To calculate effect sizes and their confidence intervals [−95%; 95%] for dependent comparisons as well as to create violin plots, the ggstatsplot package [23] was used.

Linear regression models were performed using Jamovi [24]. Changes in the body composition indicators were treated as dependent variables. Sets of behaviours related to the frequency of physical activities or product consumption were treated as predictors. Sex (being female), presence of high occupation status, age, years of education, MoCA and GDS scores, and current health status self-assessment were used as covariates in the regression models.

Alluvial diagrams and frequency tables were created using Jamovi with the easyalluvial and ClinicoPath packages [25,26].

Network graphing was performed using the qgraph package [27] in R [28]. The correlation results used for creating network plot were corrected using the False Discovery Rate (FDR) with a 0.9 cut-off.

3. Results

Overall, 205 subjects (40 males) were examined at the baseline and re-evaluated after two years. The mean age of the participants after the two year follow-up assessment was 69.67 years (−95% CI = 68.85; 95% CI = 70.5, range 60–88). The BMI of subjects was (mean ± SD) 27.54 ± 4.44 kg/m^2 during baseline vs. 27.45 ± 4.3 kg/m^2 after two years. The frequencies of BMI categories are shown in Table S1. In total, 173 participants (86%) were characterized by high occupational status. Every participant had undertaken work during at least some period of their life; therefore, there were no participants in the unemployed subgroup.

There were no significant changes in body composition within two years in the examined subjects except for the visceral fat level. No changes were observed in body weight (71.70 ± 19.03 kg during baseline vs. 70.6 ± 17 kg after 2 years, p = 0.758) (Figure 1A), nor were changes in muscle mass observed (43.55 ± 8.2 kg during baseline vs. 43.60 ± 7.1 kg after 2 years, p = 0.743) (Figure 1C). Body fat percentage did not change in a statistically significant or practically significant manner (median ± IQR (34.80 ± 10.45 % during baseline vs. 34.80 ± 9.88 % after 2 years, p = 0.713) (Figure 1D). Visceral fat level, on the other hand, was reduced significantly (10 ± 0.25 units during baseline vs. 9 ± 0.27 units after 2 years, p = 0.00000003, r = 0.55, CI 95%) (Figure 1B) [0.43, 0.66].

A relatively high variance in body mass changes as well as in its composition was noted; see Figure 1. Therefore, linear regression was applied to build a model to find significant predictors of body composition changes. Changes in body fat (%), skeletal muscle mass (kg), and visceral fat (units) within two years served as predicted variables. Table S2 shows the confounders included in the models. Tables S3 and S4 show a list of physical activities and diet product consumption frequency included in the regression models as dummy variables. Figures S1–S3 show the frequency and its change over a period of two years for physical activities (Figure S1), selected food products (Figure S2), and alcohol consumption (Figure S3). In the case of tennis, team games, sailing/riding on a horse, and skiing, the frequency of participants who were undertaking those activities was relatively small; therefore, these activities were omitted from further analysis. In addition, the number of categories was reduced to combine "never" with "once a year" and "several times per year" in one group. In the case of fish consumption, there were only two instances of the "daily" category; therefore, these were added to the "a few times per week" subgroup.

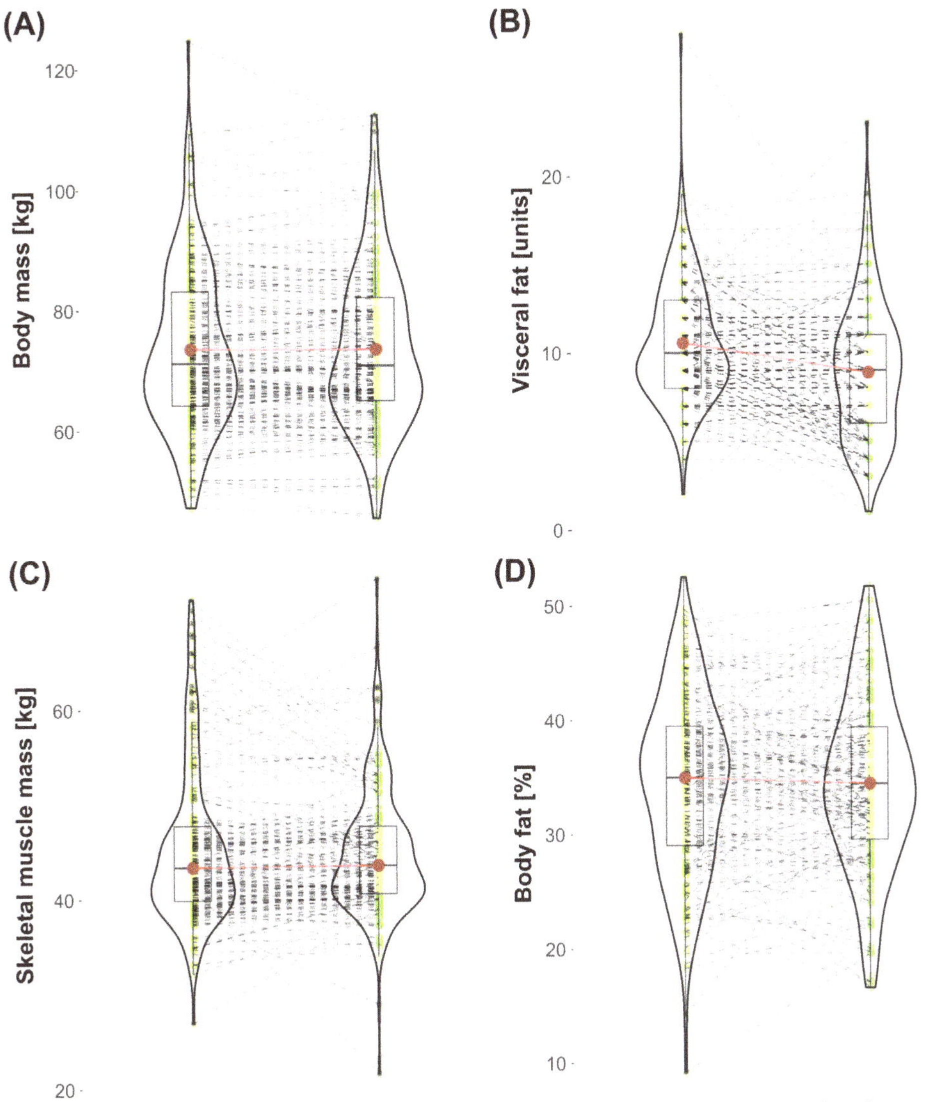

Figure 1. Changes in body mass and composition within two years: (**A**) change in body mass (kg) vs. after 2 years, (**B**) change in visceral fat (units) vs. after 2 years, (**C**) change in muscle mass (kg) vs. after 2 years, (**D**) change in body fat (%) vs. after 2 years. The shape of the violin graph corresponds to the distribution of values. The median value is illustrated by the horizontal black line inside the box, and the green dots connected by dashed grey lines show the results of individual participants. The arithmetical mean value is illustrated by the dark red dots connected by a light red line.

Table 1 presents results from a model that explained changes in body fat percentage based on food product consumption frequency (Adjusted $R^2 = 0.17$, $F = 1.91$, $p = 0.003$). Drinking beer a few times per week was associated with a significant increase in body fat percentage in comparison to subjects who drink beer "never" to "a few times per year" (an increase of 8.99%, $p = 0.013$). Vodka and other 40-proof alcohol and red wine consumption frequency were not significantly associated with body fat percent changes ($p < 0.05$). Subjects who consumed sweets a few times per month, once a week, a few times per week, or daily noted an increase in body fat percentage in comparison to subjects who consumed sweets "never" to "a few times per year" (increase in body fat of 5.41% ($p = 0.023$), 4.96% ($p = 0.021$), 4.45 % ($p = 0.038$), and 4.83% ($p = 0.035$), respectively). Participants who drank coffee a few times per week had a significantly higher increase in body fat percentage in comparison to subjects who consumed coffee daily (3.88% of body fat increase, $p = 0.029$). Subjects who consumed green or white tea "never" to "a few times per year" saw a decrease in body fat percentage in comparison to subjects who consumed green or white tea daily (−3.18 % of body fat, $p = 0.028$). Males gained 5.06% body fat in comparison to females ($p < 0.001$) (Table 1).

Table 1. Food product consumption frequencies as predictors of changes in body fat percentage within two years.

			95% Confidence Interval			
Predictor	**Estimate**	**SE**	**Lower**	**Upper**	**t**	***p***
Intercept [a]	20.34	11.91	−3.22	43.89	1.71	0.090
Sex [F]:						
0–1	5.04	1.46	2.16	7.92	3.46	<0.001
High OS [presence]:						
0–1	−0.72	1.77	−4.23	2.78	−0.41	0.684
Age [years]	−0.13	0.09	−0.31	0.04	−1.50	0.136
MoCA [pts]	−0.08	0.15	−0.37	0.21	−0.53	0.596
GDS [pts]	−0.20	0.21	−0.61	0.22	−0.95	0.345
Health status assessment currently [pts]	−0.30	0.35	−0.99	0.39	−0.86	0.390
Education [years]	−0.08	0.16	−0.39	0.24	−0.47	0.638
Drinking beer [freq]:						
3–0	0.41	1.39	−2.35	3.16	0.29	0.771
4–0	0.67	2.47	−4.21	5.56	0.27	0.786
5–0	8.99	3.59	1.89	16.08	2.51	0.013
Chocolate [freq]:						
3–0	−3.01	1.77	−6.51	0.49	−1.70	0.091
4–0	−2.15	1.76	−5.62	1.32	−1.22	0.223
5–0	−3.41	1.80	−6.98	0.15	−1.89	0.060
6–0	−0.29	2.37	−4.99	4.41	−0.12	0.903
Sweets [freq]:						
3–0	5.41	2.35	0.75	10.06	2.30	0.023
4–0	4.96	2.12	0.77	9.16	2.34	0.021
5–0	4.45	2.13	0.24	8.66	2.09	0.038
6–0	4.83	2.27	0.35	9.31	2.13	0.035
Fish [freq]:						
0–4	0.80	1.64	−2.45	4.05	0.49	0.627
3–4	−0.58	1.31	−3.17	2.01	−0.44	0.658
5–4	0.27	1.54	−2.79	3.33	0.17	0.862
Coffee [freq]:						
0–6	2.85	2.08	−1.26	6.96	1.37	0.173
3–6	2.75	2.70	−2.59	8.09	1.02	0.310
4–6	−1.29	2.88	−6.99	4.41	−0.45	0.654
5–6	3.88	1.76	0.40	7.37	2.20	0.029
Green and/or white tea [freq]:						
0–6	−3.20	1.42	−6.02	−0.39	−2.25	0.026
3–6	−1.86	1.89	−5.60	1.88	−0.98	0.328
4–6	0.44	1.95	−3.42	4.30	0.23	0.821
5–6	−0.49	1.48	−3.41	2.43	−0.33	0.738

Table 1. *Cont.*

Predictor	Estimate	SE	95% Confidence Interval		t	*p*
			Lower	Upper		
Red wine [freq]:						
3–0	−2.48	1.26	−4.97	0.01	−1.97	0.051
4–0	1.59	1.96	−2.30	5.47	0.81	0.420
5–0	−0.22	2.39	−4.95	4.51	−0.09	0.926
6–0	−0.07	5.00	−9.96	9.81	−0.01	0.988
Vegetables and fruit [freq]:						
4–3	−5.95	6.85	−19.49	7.60	−0.87	0.387
5–3	−7.49	6.57	−20.49	5.51	−1.14	0.257
6–3	−7.55	6.40	−20.21	5.11	−1.18	0.240
Vodka and other 80-proof alcohols [freq]:						
3–0	0.31	1.43	−2.52	3.13	0.21	0.831
4–0	−1.40	2.26	−5.87	3.07	−0.62	0.536
5–0	10.20	7.00	−3.64	24.05	1.46	0.147
6–0	6.58	7.07	−7.41	20.57	0.93	0.354

[a]: reference level; 0: never/once a year/several times per year; 3: 1–2 times per month; 4: once a week; 5: several times per week; 6: daily.

Table S3 includes the frequency of physical activities, while Table S4 shows the frequency of food product consumption included in the regression analysis. Table S5 presents results from a model that explained changes in muscle mass in kilograms based on food product consumption frequency (Adjusted $R^2 = 0.16$, $F = 1.83$, $p = 0.006$). Drinking beer a few times per week was related to a decrease in muscle mass in comparison to drinking beer never to a few times per year (−5.5 kg, $p = 0.043$). Sweets consumption one to two times per month was related to a decrease in muscle mass in comparison to consuming sweets never to a few times per year (−3.59 kg, $p = 0.047$). Contrarily, drinking red wine one to two times per month was related to an increase in muscle mass in comparison to drinking red wine never to a few times per year (1.88 kg, $p = 0.049$). For the rest, food product consumption frequency was not a significant predictor of changes in muscle mass within the two years (Table S2). Males lost a mean of 4.03 kg less of skeletal muscle mass in comparison to females ($p < 0.001$) (Table S5).

Table S6 presents results from a model that explained changes in units of visceral fat based on food product consumption frequency (Adjusted $R^2 = 0.08$, $F = 1.37$, $p = 0.096$). Food product consumption frequency was not a significant predictor of visceral fat within the two years (Table S6).

Tables S7–S9 present results from models that explained changes in body fat percentage (Table S7), muscle mass in kilograms (Table S8), and visceral fat in units (Table S9) in relation to the frequency of physical activities. The models were not statistically significant, nor were particular predictors related to physical activity frequency ($p > 0.05$).

Figure 2 and Tables S10–S13 show the relationship between the change in body fat percentage over two years presented as a binary value (increased vs. decreased) in comparison with the frequency of food products period denoted as significant within the same in the linear regression model. Overall, older participants who noted an increase in body fat percentage within two years seemed to drink beer more frequently during this period (Table S10); 39 % of those who noted a body fat decrease consumed beer a few times per year or less frequently, compared to 33 % of participants who noted a body fact increase (Table S10). Of patients who drank green or white tea never to several times per year, n = 21 (12% of the total sample) noted an increase in body weight (Table S11). Table S12 shows that daily consumers of coffee, n = 59 (33% of the total sample) noted an increase, while n = 79 (44% of the total sample) noted a decrease in body fat percentage within two years. Table S13 shows that in participants who ate sweets never to several times per year, 6% noted a decrease in body fat percentage, while 2% noted an increase.

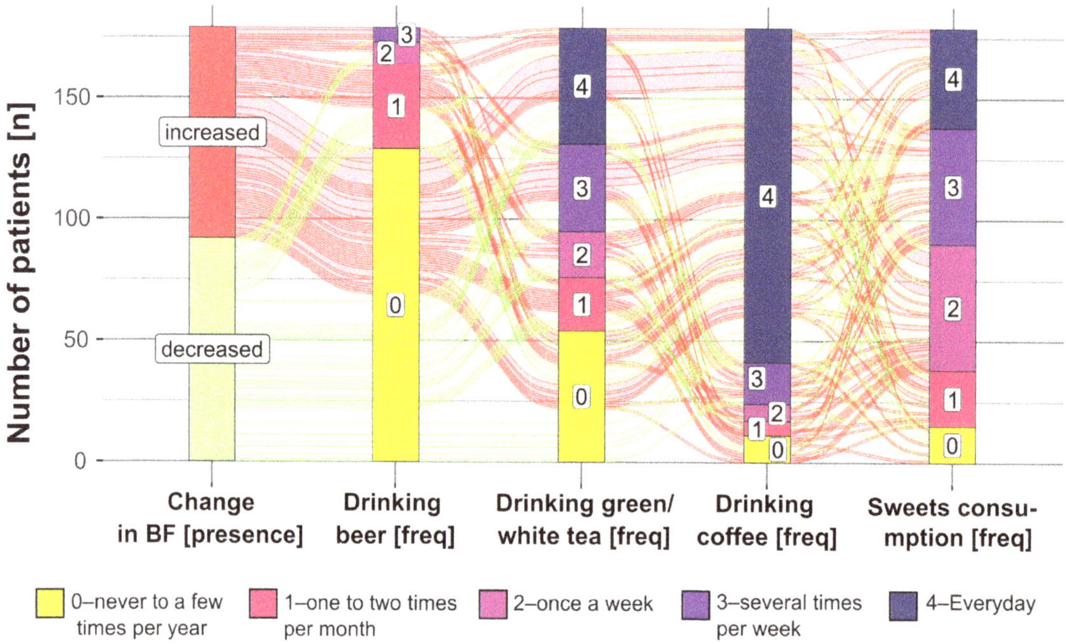

Figure 2. Alluvial diagram showing the relationship between the presence of body fat (BF) change with the frequency of selected food products consumption.

In addition, as can be seen in Figure 2, the frequencies of consumption of products related to body fat percent change within two years in this study seem to be interrelated. For instance, 28 % of participants drank green or white tea a few times per year while drinking coffee daily (Table S14). Contrarily, 22 % of participants drank both beverages daily (Table S14). In addition, there seems to be a positive relationship between sweets and coffee consumption; subjects who ate sweets the most frequently (once a week or more frequently) more often consumed coffee daily (Table S15). For instance, 6 % of participants ate sweets a few times per year and drank coffee daily; however, 21%, 22%, and 21% of participants drank coffee daily and consumed sweets once a week, a few times per week, or daily, respectively (Table S15).

As a part of the exploratory approach, we performed a network analysis between the examined factors (Figure 3). Interestingly, the change in the percentage of body fat was negatively related to the change in skeletal muscle mass. The created network confirmed the presence of multiple intra-group relationships meaning that there were many positive relationships between the frequency of consumption of particular food products and other food products as well as between particular physical activities and other physical activities. Interestingly, after FDR, only three edges were indicating inter-group relations between food consumption and physical activity frequency: the frequency of consumption of green tea was positively related to short walks, and the frequency of sweets consumption was negatively related to the frequency of running/jogging. A negative relationship between the frequency of drinking vodka and other 80-proof alcohols with running/jogging frequency was noted, in that a higher frequency of strong alcohol consumption was associated with a lower frequency of running (Figure 3). In addition, high occupational status was related to more frequent running/jogging, Nordic walking, and consumption of red wine (Figure 3), while being female was related to a lower frequency of beer consumption.

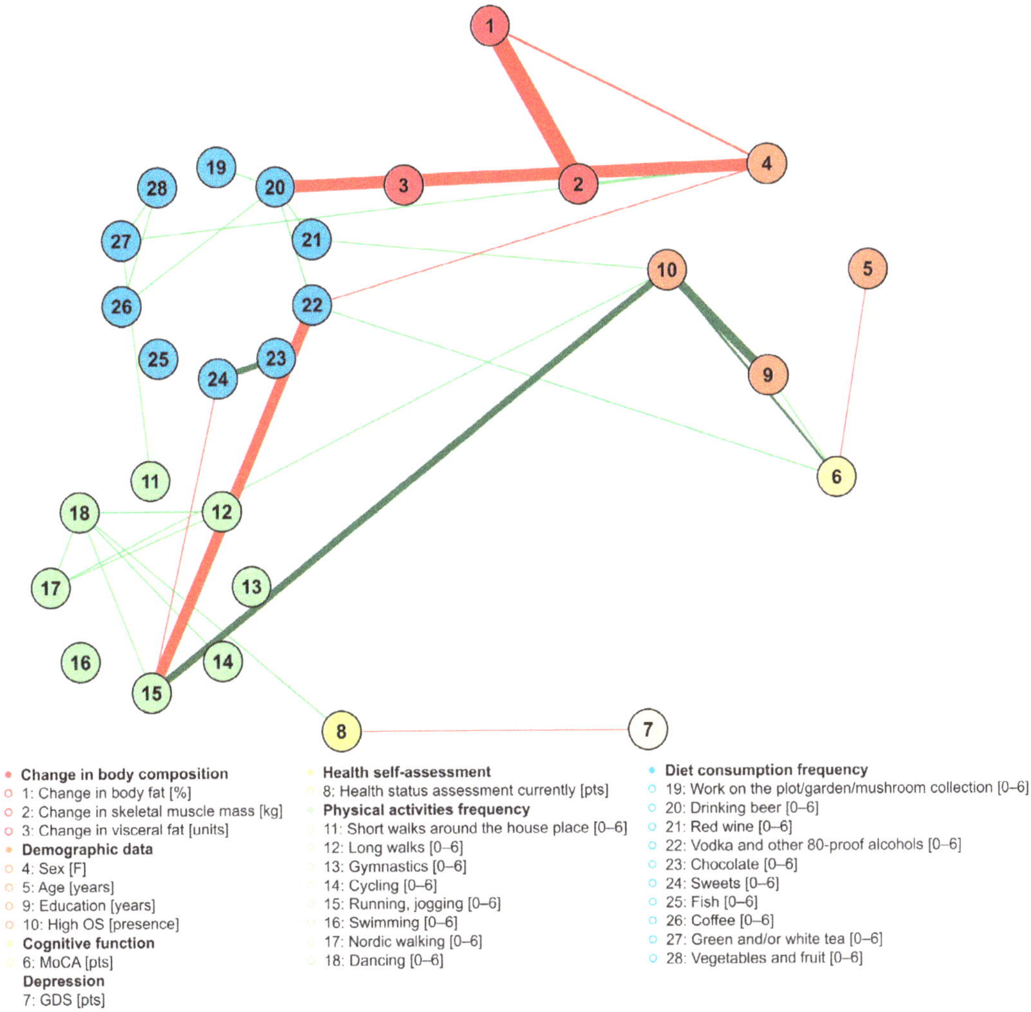

Figure 3. Network analysis of the relationship of body composition change with the frequency of physical activities, food products consumption, demographic data, cognitive function, depression severity, and health self-assessment. Nodes were grouped according to pre-specified groups, to which each variable belongs. Illustrated graphs are weighted, with green edges showing positive weights and red edges showing negative weights. The color saturation and the width of the edges are proportional to the absolute weight and scale relative to the strongest weight in the graph. OS—occupational status; MoCA—Montreal Cognitive Assessment; GDS—Geriatric Depression Scale.

4. Discussion

In the current study, no significant changes in body weight, muscle mass, or body fat percentage were observed. However, a significant reduction in visceral fat level was noted. Changes in all body composition measurements observed within two years were characterized by relatively high variance. The frequencies of food products consumption are interrelated. For instance, subjects who ate sweets the most frequently (once a week or more frequently) consumed coffee daily more often. Such a network relationship between particular behaviours was observed in previous studies as well. For instance, Mattioli et al.

noted good adherence to the Mediterranean diet and high levels of physical activity in a group of women with high coffee consumption [29].

4.1. Body Composition Trajectory in Older People

As body composition has a significant impact on disease development and physical dependence, there is great interest in understanding the progression of body composition changes to help prevent or at least mitigate them [30]. Numerous studies have evaluated or determined this progression of body composition changes, providing important results. For example, a study conducted by Visser et al. [31] aimed to assess changes in body composition in healthy older people over 60 years of age with a two-year follow-up, finding that older adults had an average change in body composition of 1–2% of total body mass (a decrease of 0.3% in men and 0.4% in women was reported), fat-free mass, appendicular lean soft tissue mass (ALST), and total fat mass after the two-year follow-up. There was a decrease in fat-free mass of 1.1% and in appendicular soft tissue mass (ASLT) of 0.8% in men, which was accompanied by an increase in total fat mass of 2.0%, while in women a decrease in body mass of 0.6% was reported after two years, with no change in ALST or body fat mass. Thus, the results of their study showed that men have a clear loss of ASLT compared to women and that men are at higher risk of loss of muscle function compared to women. Another longitudinal study conducted by Raguso et al. [30] aimed to assess body composition changes in people over 65 years of age (74 healthy men and 66 women) with a three-year follow-up period. A decrease in soft tissue lean tissue (FFST) and appendicular skeletal muscle mass (-0.3 ± 1.4 and -0.2 ± 2.2 kg, respectively) and an increase in body fat (0.6 ± 2.2 kg) were observed.

4.2. Physical Activities and Product Consumption Frequency as Predictors of Skeletal Muscle Mass

In the current study, drinking beer a few times per week was related to a decrease in muscle mass in comparison to drinking beer never to a few times per year. Sweets consumption one to two times per month was related to a decrease in muscle mass in comparison to consuming sweets never to a few times per year. Contrarily, drinking red wine one to two times per month was related to an increase in muscle mass in comparison to drinking red wine never to a few times per year. The frequency of physical activities was not related to changes in body composition. Somewhat contrary to the results observed in the current study, higher muscle mass was related to a lower percentage of total and trunk fat in individuals who are [30]. Indeed, the role of physical activity and diet in maintaining lean body mass in older subjects has been underlined numerous times in the literature [32]. A recent meta-analysis showed that a resistance exercise training program was effective in inducing skeletal muscle hypertrophy in subjects 75 years of age and older [33].

4.3. Effects of Food Product Consumption Frequency on Changes in Body Composition within Two Years

In the current study, more frequent drinking of beer and of green or white tea as well as sweets consumption were related to a higher increase in body fat percentage within two years. Daily coffee consumption was related to a lower increase in body fat percentage in comparison to consumers who drank coffee a few times per week. More frequent drinking of beer and consumption of sweets was related to a decrease in skeletal muscle mass over two years. Contrarily, drinking red wine one to two times per month was related to an increase in muscle mass in comparison to drinking red wine never to a few times per year.

In the current study, it was observed that consuming sweets a few times per month, once a week, a few times per week, or daily was related to an increase in body fat percentage in comparison to consuming sweets never up to a few times per year. In the current paradigm on weight and body composition management, it seems that the balance between caloric intake and output is the most important factor. For obese people, daily calorie restriction is a tried-and-true primary weight-loss method [34]. No significant differences have been shown in weight loss induced by a year-long low-fat diet vs. a low-carbohydrate isocaloric diet [35]. If the results of the above-described study can be replicated, then a lack of practical differences between the quality of diets with the same caloric deficit might lead to the development of a very flexible approach to obesity treatment. Presumably, if patients were allowed to choose their diet type, this could be related to specific product abundance (for instance, a relatively higher amount of carbohydrate-rich food in a low-fat diet). Such an approach might eventually lead to higher adherence to the diet, thereby increasing its efficacy.

However, it seems that the intake the certain food products and their consequences can indirectly influence the above-mentioned balance. In the current study, participants who drank coffee a few times per week had a higher increase in body fat percentage in comparison to subjects who consumed coffee daily. Subjects who consumed green or white tea never to a few times per year saw a decrease in body fat percentage in comparison to subjects who consumed green or white tea daily. Polyphenols in green tea might lead to a reduction in chronic inflammation of the liver and gastrointestinal tract and could change the gastrointestinal microbiota [36]. Modulation of the gut microbiota might in turn be related to body composition changes [37]. Contrary to our observation, tea and its components have been shown to decrease both body fat stores and body mass [36]. Presumably, there are independent mechanisms of epigallocatechin and caffeine from tea leaves that have synergistic effects on weight loss [36]. In line with our results, it has been shown that ingestion of coffee might lead to a decrease in storing of fat in the body due to inhibition of the multiplication of adipocytes, modulation of the activity of transcription factors taking part in lipid production, and the alternation of gut microbiota [36]. Nevertheless, two things should be noted about the results of the current study. First, the above-mentioned relationships between the frequency of tea and coffee consumption and body composition changes were characterized by a small effect size. Second, these relationships might in fact be spurious ones. It seems unlikely that the consumption of green or white tea or coffee has a physiologically significant effect on body composition per se. Presumably, the effects of the frequency of consumed products on body composition changes might be indirect, through modulation of gut microbiota and other mechanisms, which would eventually be related to modulation of kilocalories intake and/or expenditure, and in turn to changes in body composition. The results of previous studies have shown that decreased carbohydrate availability due to fasting or a ketogenic diet (KD), metabolic consequences of an intense physical exercise session, or impaired insulin signaling might lead to increased production of ketone bodies (KBs) [38]. KDs appear to reduce appetite, which in turn leads to a decrease in kilocalorie intake [39–42]. In addition, it has been proposed that KBs could lead to the modulation of circadian rhythms, including appetite, sleep, and hormone release [43]. Nevertheless, more studies on the potential anorexigenic effect of KD and associated mechanisms are needed [43]. Because of the potential risk of KD and limited evidence on its application in type 2 diabetes, caution in the application of KD has been advised [44]. Therefore, further studies should assess the differences between different types of diets (i.e., those low in carbohydrates vs. those relatively high in carbohydrates) with the same caloric deficit.

4.4. Study Limitations

In the current study, only self-reported recall of the frequency of food product consumption and physical activity was measured, without any indication of the quantity or quality. Future longitudinal studies conducted in Poland should measure more factors related to participants' behaviour.

In the current study, we used a body composition measurement based on a bioelectrical impedance device. This technology has several limitations in estimating body composition with high accuracy [43]. In further studies, it would be best to accompany this method with body composition and structure measurement using dual-energy X-Ray absorptiometry or magnetic resonance imaging and waist-to-hip ratio.

Further studies should be conducted on a larger sample size and should allow for adding more confounding factors into models, including effects of pathology, diagnosis, and treatment of chronic disorders. Characteristics of diet should be examined (plant vs. animal origin). Moreover, a larger sample might lead to higher frequencies in particular categories of qualitative variables, leading to more balanced data. For instance, it is a high need to examine the effects of unemployment and types of occupation related to no or restricted opportunities to consume regular meals during shifts, as these could be indirectly related to prolonged intermittent fasting during work time. As the R^2 values of created models are notably small in the current study, further studies should extend the list of examined behaviors. In addition, the effect sizes of relationships between behaviors of older participants with body composition change should be provided together with a discussion on the practical and clinical significance of the results. It should be noted that behaviors denoted in the current study as statistically significant in relation to body composition change in older people may not necessarily be practically meaningful due to potential low effect sizes and/or problems with adherence of the subjects to interventions.

As we mentioned in our previous study [19], we did not measure the method of coffee preparation in the currently analysed sample. The method used for coffee preparation might affect both the quality and quantity of substances introduced upon consumption [27,45]. Nevertheless, based on population data available in Poland, it can be predicted that the vast majority of coffee was prepared using espresso prepared under pressure in a machine or by pouring hot water over coffee (coffee in a glass or cup, without filtering, with coffee grounds) [46].

As highlighted in one of our previous studies, there is an urgent need for further longitudinal studies and the incorporation of a more representative sample with a higher proportion of males [19].

5. Conclusions

In the current longitudinal study, in which older participants were examined during the baseline and after two years, we have observed the following:

1. No significant changes in body weight, muscle mass, or body fat percentage were observed.
2. A significant reduction in visceral fat level was noted. Changes in all body composition measurements were characterized by relatively high variance.
3. More frequent drinking of beer, drinking of green or white tea, and consumption of sweets were related to a higher increase in body fat percentage within two years. Daily coffee consumption was related to a lower increase in body fat percentage in comparison to consumers who drank coffee a few times per week.
4. More frequent drinking of beer and consumption of sweets was related to a decrease in skeletal muscle mass within two years. Contrarily, drinking red wine one to two times per month was related to an increase in muscle mass in comparison to drinking red wine never to a few times per year.
5. Drinking beer a few times per week was related to a decrease in muscle mass in comparison to drinking beer never to a few times per year. Consuming sweets one to two times per month was related to a decrease in muscle mass in comparison to consuming sweets never to a few times per year. Contrarily, drinking red wine one to two times per month was related to an increase in muscle mass in comparison to drinking red wine never to a few times per year
6. Frequency of physical activities was not related to changes in body composition.
7. The frequencies of food product consumption are interrelated; for instance, subjects who ate sweets the most frequently (once a week or more frequently) were more often daily consumers of coffee.

Supplementary Materials: The following supporting information can be downloaded at: https://www.mdpi.com/article/10.3390/nu15081834/s1, Figure S1. Bar plots showing the relative frequency of physical activities before and after two years, Figure S2. Bar plots showing the relative frequency of selected products consumption before and after two years, Figure S3. Bar plots showing the relative frequency of alcohol consumption before and after two years, Table S1. Frequencies of BMI categories before and after two years, Table S2. List of confounding factors included in the linear regression models, Table S3. List of physical activities frequency included in the regression analysis, Table S4. List of food products consumption frequencies included in the regression analysis, Table S5. Food products consumption frequency as predictors of changes in muscle mass in kilograms within 2 years, Table S6. Food products consumption frequency as predictors of changes in visceral fat in units within 2 years, Table S7. Physical activities frequency as predictors of changes in body fat percent within 2 years, Table S8. Physical activities frequency as predictors of changes in muscle mass in kilograms within 2 years, Table S9. Physical activities frequency as predictors of changes in visceral fat in units within 2 years, Table S10. Relationship between frequency of drinking beer to the increased vs. decreased body fat, Table S11. Relationship between frequency of drinking green or white tea to the increased vs. decreased body fat, Table S12. Relationship between frequency of drinking coffee to the increased vs. decreased body fat, Table S13. Relationship between frequency of sweets consumption to the increased vs. decreased body fat, Table S14. Relationship between frequency of green/white tea and coffee consumption, Table S15. Relationship between frequency coffee and sweets consumption frequency.

Author Contributions: Conceptualization, A.K. and S.K.; methodology, A.K., S.K., P.Z. and K.K.-K.; software, S.K.; validation, A.K.; formal analysis, S.K.; investigation, S.K. and A.K.; resources, P.Z. and K.K.-K.; data curation, A.K. and S.K.; writing—original draft preparation, A.K. and S.K.; writing—review and editing, A.K., G.F.L.S., F.H., S.K., P.Z. and K.K.-K.; visualization, S.K.; supervision, P.Z. and K.K.-K.; project administration, A.K., S.K., P.Z. and K.K.-K.; funding acquisition, K.K.-K. All authors have read and agreed to the published version of the manuscript.

Funding: This research received no external funding. Dr. Guillermo F. López Sánchez is funded by the European Union–Next Generation EU.

Institutional Review Board Statement: The study was conducted according to the guidelines of the Declaration of Helsinki and approved by the ethics committee of Ludwik Rydygier Memorial Collegium Medicum in Bydgoszcz, Nicolaus Copernicus University, Torun (KB 340/2015, date of approval: 21 April 2015).

Informed Consent Statement: Informed consent was obtained from all subjects involved in the study.

Data Availability Statement: Individual data are available from author S.K. upon request.

Acknowledgments: We would like to give many thanks to Karl J. Morten, Nuffield Department of Women's and Reproductive Health, University of Oxford, Oxford, UK, for his help in proofing this manuscript.

Conflicts of Interest: The authors declare no conflict of interest.

References

1. Franceschi, C.; Garagnani, P.; Morsiani, C.; Conte, M.; Santoro, A.; Grignolio, A.; Monti, D.; Capri, M.; Salvioli, S. The continuum of aging and age-related diseases: Common mechanisms but different rates. *Front. Med.* **2018**, *5*, 61. [CrossRef]
2. Peralta, M.; Ramos, M.; Lipert, A.; Martins, J.; Marques, A. Prevalence and trends of overweight and obesity in older adults from 10 European countries from 2005 to 2013. *Scand. J. Public Health* **2018**, *46*, 522–529. [CrossRef] [PubMed]
3. Flatt, T.; Partridge, L. Horizons in the evolution of aging. *BMC Biol.* **2018**, *16*, 93. [CrossRef] [PubMed]
4. Kennedy, B.K.; Berger, S.L.; Brunet, A.; Campisi, J.; Cuervo, A.M.; Epel, E.S.; Franceschi, C.; Lithgow, G.J.; Morimoto, R.I.; Pessin, J.E.; et al. Geroscience: Linking aging to chronic disease. *Cell* **2014**, *159*, 709–713. [CrossRef] [PubMed]
5. Al Snih, S.; Ottenbacher, K.J.; Markides, K.S.; Kuo, Y.F.; Eschbach, K.; Goodwin, J.S. The effect of obesity on disability vs mortality in older Americans. *Arch. Intern. Med.* **2007**, *167*, 774–780. [CrossRef]
6. Barzilay, J.I.; Blaum, C.; Moore, T.; Xue, Q.L.; Hirsch, C.H.; Walston, J.D.; Fried, L.P. Insulin resistance and inflammation as precursors of frailty: The cardiovascular health study. *Arch. Intern. Med.* **2007**, *167*, 635–641. [CrossRef] [PubMed]

7. Frisardi, V.; Solfrizzi, V.; Seripa, D.; Capurso, C.; Santamato, A.; Sancarlo, D.; Vendemiale, G.; Pilotto, A.; Panza, F. Metabolic-cognitive syndrome: Across-talk between metabolic syndrome and Alzheimer's disease. *Ageing Res. Rev.* **2010**, *9*, 399–417. [CrossRef]
8. Cevenini, E.; Monti, D.; Franceschi, C. Inflammageing. *Curr. Opin. Clin. Nutr. Metab. Care* **2013**, *16*, 14–20. [CrossRef] [PubMed]
9. Leenders, M.; Verdijk, L.; van der Hoeven, L.; Adam, J.J.; van Kranenburg, J.; Nilwik, R.; van Loon, L.J. Patients with type 2diabetes show a greater decline in muscle mass, muscle strength, and functional capacity with aging. *J. Am. Med. Dir. Assoc.* **2013**, *14*, 585–592. [CrossRef]
10. Yamauchi, T.; Midorikawa, T.; Hagihara, J.; Sasaki, K. Quality of life nutritional status, physical activity, and their interrelationships of elderly living on an underpopulated island in Japan. *Geriatr. Gerontol. Int.* **2007**, *7*, 26–33. [CrossRef]
11. Hu, Z.; Qin, L.; Kaminga, A.C.; Xu, H. Relationship between multiple lifestyle behaviors and health-related quality of life among elderly individuals with prediabetes in rural communities in China: A STROBE-compliant article. *Medicine* **2020**, *99*, e19560. [CrossRef] [PubMed]
12. Prochaska, J.J.; Spring, B.; Nigg, C.R. Multiple health behavior change research: An introduction and overview. *Prev. Med.* **2008**, *46*, 181–188. [CrossRef]
13. Mkhitaryan, S.; Crutzen, R.; Steenaart, E.; de Vries, N.K. Network approach in health behavior research: How can we explore new questions? *Health Psychol. Behav. Med.* **2019**, *7*, 362–384. [CrossRef] [PubMed]
14. Ahmed, W.; Vidal-Alaball, J.; Lopez Segui, F.; Moreno-Sánchez, P.A. A Social Network Analysis of Tweets Related to Masks during the COVID-19 Pandemic. *Int. J. Environ. Res. Public Health* **2020**, *17*, 8235. [CrossRef] [PubMed]
15. Hevey, D. Network analysis: A brief overview and tutorial. *Health Psychol. Behav. Med.* **2018**, *6*, 301–328. [CrossRef] [PubMed]
16. Rom, O.; Reznick, A.Z.; Keidar, Z.; Karkabi, K.; Aizenbud, D. Body composition in heavy smokers: Comparison of segmental bioelectrical impedance analysis and dual-energy X-ray absorptiometry. *Adv. Exp. Med. Biol.* **2015**, *840*, 1–11. [CrossRef]
17. Stahn, A.; Terblanche, E.; Gunga, H.-C. Use of bioelectrical impedance: General principles and overview. In *Handbook of Anthropometry: Physical Measures of Human form in Health and Disease*; Preedy, V.R., Ed.; Springer: New York, NY, USA, 2012; pp. 49–90.
18. Bailey, K.V.; Ferro-Luzzi, A. Use of body mass index of adults in assessing individual and community nutritional status. *Bull. World Health Organ.* **1995**, *73*, 673–680.
19. Kujawski, S.; Kujawska, A.; Gajos, M.; Topka, W.; Perkowski, R.; Androsiuk-Perkowska, J.; Newton, J.L.; Zalewski, P.; Kędziora-Kornatowska, K. Cognitive functioning in older people. results of the first wave of cognition of older people, education, recreational activities, nutrition, comorbidities, and functional capacity studies (COPERNICUS). *Front. Aging Neurosci.* **2018**, *10*, 421. [CrossRef]
20. Almeida, O.P.; Almeida, S.A. Short versions of the geriatric depression scale: A study of their validity for the diagnosis of a major depressive episode according to ICD-10 and DSM-IV. *Int. J. Geriatr. Psychiatry* **1999**, *14*, 858–865. [CrossRef]
21. Conradsson, M.; Rosendahl, E.; Littbrand, H.; Gustafson, Y.; Olofsson, B.; Lövheim, H. Usefulness of the Geriatric Depression Scale 15-item version among very old people with and without cognitive impairment. *Aging Ment. Health* **2013**, *17*, 638–645. [CrossRef]
22. Nasreddine, Z.S.; Phillips, N.A.; Bédirian, V.; Charbonneau, S.; Whitehead, V.; Collin, I.; Cummings, J.L.; Chertkow, H. The Montreal Cognitive Assessment; MoCA: A brief screening tool for mild cognitive impairment. *J. Am. Geriat Soc.* **2005**, *53*, 695–699. [CrossRef]
23. Patil, I. Ggstatsplot: 'ggplot2' Based Plots with Statistical Details. CRAN. 2018. Available online: https://mran.revolutionanalytics.com/snapshot/2018-09-28/web/packages/ggstatsplot/ggstatsplot.pdf (accessed on 13 February 2023).
24. The Jamovi Project. Jamovi (Version 2.3). [Computer Software]. 2022. Available online: https://www.jamovi.org (accessed on 13 February 2023).
25. Bjoern Koneswarakantha. Easyalluvial: Generate Alluvial Plots with a Single Line of Code. [R Package]. 2019. Available online: https://CRAN.R-project.org/package=easyalluvial (accessed on 13 February 2023).
26. Serdar Balci. ClinicoPath Jamovi Module. [R Package]. 2022. Available online: https://github.com/sbalci/ClinicoPathJamoviModul (accessed on 13 February 2023).
27. Epskamp, S.; Cramer, A.O.J.; Waldorp, L.J.; Schmittmann, V.D.; Borsboom, D. qgraph: Network Visualizations of Relationships in Psychometric Data. *J. Stat. Softw.* **2012**, *48*, 1–18. [CrossRef]
28. R Core Team. *R: A Language and Environment for Statistical Computing*; R Foundation for Statistical Computing: Vienna, Austria, 2021. Available online: https://www.R-project.org/ (accessed on 13 February 2023).
29. Mattioli, A.V. Effects of caffeine and coffee consumption on cardiovascular disease and risk factors. *Future Cardiol.* **2007**, *3*, 203–212. [CrossRef] [PubMed]
30. Raguso, C.A.; Kyle, U.; Kossovsky, M.P.; Roynette, C.; Paoloni-Giacobino, A.; Hans, D.; Genton, L.; Pichard, C. A 3-year longitudinal study on body composition changes in the elderly: Role of physical exercise. *Clin. Nutr.* **2006**, *25*, 573–580. [CrossRef]
31. Visser, M.; Pahor, M.; Tylavsky, F.; Kritchevsky, S.B.; Cauley, J.A.; Newman, A.B.; Blunt, B.A.; Harris, T.B. One- and two-year change in body composition as measured by DXA in a population-based cohort of older men and women. *J. Appl. Physiol.* **2003**, *94*, 2368–2374. [CrossRef] [PubMed]

32. Chopp-Hurley, J.N.; Wiebenga, E.G.; Keller, H.H.; Maly, M.R. Nutrition risk, physical activity and fibre intake are associated with body composition in OA: Analysis of the Canadian Longitudinal Study on Aging. *BMJ Nutr. Prev. Health* **2022**, *5*, 191–200. [CrossRef]
33. Grgic, J.; Garofolini, A.; Orazem, J.; Sabol, F.; Schoenfeld, B.J.; Pedisic, Z. Effects of resistance training on muscle size and strength in very elderly adults: A systematic review and meta-analysis of randomized controlled trials. *Sports Med.* **2020**, *50*, 1983–1999. [CrossRef]
34. Jensen, M.D.; Ryan, D.H.; Apovian, C.M.; Ard, J.D.; Comuzzie, A.G.; Donato, K.A.; Hu, F.B.; Hubbard, V.S.; Jakicic, J.M.; Kushner, R.F.; et al. 2013 AHA/ACC/TOS guideline for the management of overweight and obesity in adults: A report of the American College of Cardiology/American Heart Association Task Force on Practice Guidelines and the Obesity Society. *Circulation* **2014**, *129* (Suppl. 2), S102–S138. [CrossRef]
35. Gardner, C.D.; Trepanowski, J.F.; Del Gobbo, L.C.; Hauser, M.E.; Rigdon, J.; Ioannidis, J.P.; Desai, M.; King, A.C. Effect of low-fat vs low-carbohydrate diet on 12-month weight loss in overweight adults and the association with genotype pattern or insulin secretion: The DIETFITS randomized clinical trial. *JAMA* **2018**, *319*, 667–679. [CrossRef]
36. Sirotkin, A.V.; Kolesarova, A. The anti-obesity and health-promoting effects of tea and coffee. *Physiol. Res.* **2021**, *70*, 161–168. [CrossRef] [PubMed]
37. Muscogiuri, G.; Cantone, E.; Cassarano, S.; Tuccinardi, D.; Barrea, L.; Savastano, S.; Colao, A. Obesity Programs of nutrition, Education, Research and Assessment (OPERA) group. Gut microbiota: A new path to treat obesity. *Int. J. Obes. Suppl.* **2019**, *9*, 10–19. [CrossRef]
38. Fukao, T.; Lopaschuk, G.D.; Mitchell, G.A. Pathways and control of ketone body metabolism: On the fringe of lipid biochemistry. *Prostaglandins Leukot. Essent. Fat. Acids* **2004**, *70*, 243–251. [CrossRef] [PubMed]
39. Gibson, A.A.; Seimon, R.V.; Lee, C.M.Y.; Ayre, J.; Franklin, J.; Markovic, T.P.; Caterson, I.D.; Sainsbury, A. Do ketogenic diets really suppress appetite? A systematic review and meta-analysis. *Obes. Rev.* **2015**, *16*, 64–76. [CrossRef] [PubMed]
40. Paoli, A.; Bosco, G.; Camporesi, E.M.; Mangar, D. Ketosis, ketogenic diet and food intake control: A complex relationship. *Front. Psychol.* **2015**, *6*, 27. [CrossRef] [PubMed]
41. Johnstone, A.M.; Horgan, G.W.; Murison, S.D.; Bremner, D.M.; Lobley, G.E. Effects of a high-protein ketogenic diet on hunger, appetite, and weight loss in obese men feeding ad libitum. *Am. J. Clin. Nutr.* **2008**, *87*, 44–55. [CrossRef] [PubMed]
42. Masi, D.; Spoltore, M.E.; Rossetti, R.; Watanabe, M.; Tozzi, R.; Caputi, A.; Risi, R.; Balena, A.; Gandini, O.; Mariani, S.; et al. The Influence of Ketone Bodies on Circadian Processes Regarding Appetite, Sleep and Hormone Release: A Systematic Review of the Literature. *Nutrients* **2022**, *14*, 1410. [CrossRef] [PubMed]
43. Borga, M.; West, J.; Bell, J.D.; Harvey, N.C.; Romu, T.; Heymsfield, S.B.; Dahlqvist Leinhard, O. Advanced body composition assessment: From body mass index to body composition profiling. *J. Investig. Med.* **2018**, *66*, 1–9. [CrossRef] [PubMed]
44. Joshi, S.; Ostfeld, R.J.; McMacken, M. The ketogenic diet for obesity and diabetes—Enthusiasm outpaces evidence. *JAMA Intern. Med.* **2019**, *179*, 1163–1164. [CrossRef]
45. Bae, J.H.; Park, J.H.; Im, S.S.; Song, D.K. Coffee and health. *Integr. Med. Res.* **2014**, *3*, 189–191. [CrossRef]
46. Przybysz, M.A.; Widla, G.; Dluzewska, E. Preferencje konsumenckie picia kawy. Wpływ temperatury i czasu prażenia ziarna kawy na ocenę smaku i zapachu espresso. *Zesz. Probl. Postępów Nauk. Rol.* **2013**, *572*, 65–79.

nutrients

MDPI

Review

Resveratrol, a Multitasking Molecule That Improves Skeletal Muscle Health

Luana Toniolo [1,*], Monica Concato [2] and Emiliana Giacomello [2,*]

1 Laboratory of Muscle Biophysics, Department of Biomedical Sciences, University of Padova, 35131 Padova, Italy
2 Department of Medicine, Surgery and Health Sciences, University of Trieste, 34149 Trieste, Italy; monica.concato@studenti.units.it
* Correspondence: luana.toniolo@unipd.it (L.T.); egiacomello@units.it (E.G.)

Abstract: Resveratrol is a natural polyphenol utilized in Chinese traditional medicine and thought to be one of the determinants of the "French Paradox". More recently, some groups evidenced its properties as a calorie-restriction mimetic, suggesting that its action passes through the modulation of skeletal muscle metabolism. Accordingly, the number of studies reporting the beneficial effects of resveratrol on skeletal muscle form and function, in both experimental models and humans, is steadily increasing. Although studies on animal models confer to resveratrol a good potential to ameliorate skeletal muscle structure, function and performance, clinical trials still do not provide clear-cut information. Here, we first summarize the effects of resveratrol on the distinct components of the skeletal muscle, such as myofibers, the neuromuscular junction, tendons, connective sheaths and the capillary bed. Second, we review clinical trials focused on the analysis of skeletal muscle parameters. We suggest that the heterogeneity in the response to resveratrol in humans could depend on sample characteristics, treatment modalities and parameters analyzed; as well, this heterogeneity could possibly reside in the complexity of skeletal muscle physiology. A systematic programming of treatment protocols and analyses could be helpful to obtain consistent results in clinical trials involving resveratrol administration.

Keywords: resveratrol; skeletal muscle; clinical trials

1. Introduction

Resveratrol (RES) is a natural polyphenol found in grapes, berries and peanuts (Figure 1). Utilized in Chinese traditional medicine, in more recent times, RES is thought to be one of the determinants of the so-called "French Paradox" [1]. This is an epidemiological observation that evidences a low mortality rate from cardiovascular diseases among people in France despite their diet rich in saturated fats [2]. At the beginning of this century, after the demonstration that RES can prolong the lifespan in several organisms by interacting with the Silent Mating Type Information Regulator 1 (SIRT-1) pathway [3–5], some groups evidenced its properties as an exercise and calorie-restriction mimetic [6]. After extensive research, it is now suggested that RES's action is the result of the modulation of skeletal muscle cell metabolism and the improvement of mitochondrial quantity and quality, insulin sensitivity and motor function [7,8].

RES has been demonstrated to perform multiple functions at the level of the skeletal muscle, such as the modulation of cell metabolism, inhibition of protein catabolism and protection against cellular stress, thanks to its ability to interact with several signaling pathways [9,10]. Concisely, RES has been demonstrated to exert its metabolic effects mainly via the activation of adenosine monophosphate (AMP)-activated protein kinase (AMPK) and the consequent modulation of a signaling cascade involving SIRT-1, forkhead box O1 (FoxO1), nuclear factor erythroid-2-related factor 2 (Nrf2) and other effectors [10]. Successively, RES has been shown to regulate cyclooxygenases [11], to directly or indirectly scavenge reactive oxygen species (ROS) and nitrogen reactive species (RNS) [10] and to act as a phytoestrogen [12–14].

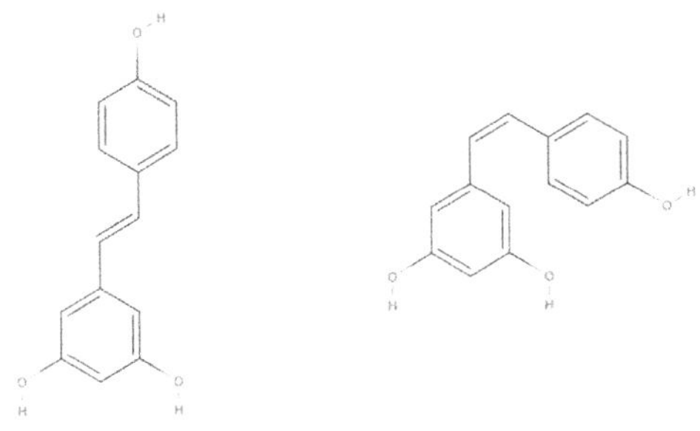

Figure 1. Structures of trans-resveratrol and cis-resveratrol.

Although studies exploiting experimental models ascribe to RES a good potential to ameliorate skeletal muscle structure, function and performance [15,16], trials on humans still do not provide clear-cut information. Actually, in addition to the heterogeneity of protocols (sample number, dose and length of treatment, age, sex and health conditions), it should be considered that skeletal muscle function is achieved thanks to the orchestration of its distinct components, such as the neuromuscular junction (NMJ), muscle fibers, connective sheaths, tendons and vessels, which could be differently influenced by RES administration [17].

The present manuscript will provide an overview of the experimental evidence supporting RES action in the distinct components of the skeletal muscle and will summarize and discuss the recent literature on human trials that investigate the effects of RES on skeletal muscle form and function.

2. Resveratrol Can Affect Distinct Skeletal Muscle Components

Skeletal muscle contraction is initiated by the nervous system with the generation of a signal that travels through motor neurons to the NMJ, inducing the release of acetylcholine, which binds to receptors on the sarcolemma of the muscle fiber. This action starts a process, the excitation–contraction coupling, that leads to the filaments sliding and muscle contracting. Skeletal muscles are attached to bones and other structures via tendons, allowing movement and respiration. Although the main component of the skeletal muscle is represented by myofibers, an optimal physical performance depends on the integration of several histological and anatomical structures with different roles in skeletal muscle contraction. Attesting to the need for efficient coordination between myofibers and associated tissues to achieve optimal skeletal muscle function, studies on aging and muscle unloading demonstrate that the loss of muscle strength is considerably greater compared to the associated alteration to the muscle mass [18–20]. This divergence could be ascribable to modifications to the NMJ, connective sheets, tendons and blood vessels, which work together to guarantee optimal contraction performance [17]. Consequently, the efficient improvement of skeletal muscle function in different physio-pathological conditions requires a strategy capable of targeting multiple muscle structures [17]. In this context, RES, initially reported to modulate the form and function of the skeletal muscle cell [9], has been demonstrated to be a pleiotropic molecule able to interact with different muscle structures, as summarized below.

2.1. Resveratrol and Skeletal Muscle Fiber

RES has been reported to act on myofibers by modulating metabolism, catabolism and oxidative stress. As largely reported, RES interacts with the AMPK-SIRT-1 pathway to trigger several signaling pathways, which induces a general myofiber remodeling similar to that seen by exercise training and calorie restriction.

RES induces the expression of genes involved in mitochondrial biogenesis and oxidative phosphorylation through the peroxisome proliferator-activated receptor gamma coactivator-alpha (PGC-1-alpha) [7]. As a result, muscle fibers from treated animals have better oxidative profiles and/or present more oxidative type I fibers, which are resistant to fatigue [7,21]. In turn, mitochondrial activity and PGC-1-alpha modulation have been demonstrated to be strictly associated with better fatty acid oxidation and lipid metabolism [7].

RES has been shown to modulate glucose metabolism by improving glucose uptake in several experimental models. As a result, mice on a high-fat diet supplemented with RES have lower circulating levels of insulin and improved glucose tolerance compared to the control group [7,8], diabetic rats display a better glucose tolerance upon RES treatment [22,23] and, more interestingly, preliminary studies confer to RES potential as an antidiabetic molecule in humans [24].

Despite the presence in the literature of conflicting data, it is suggested that RES may prevent muscle wasting in different conditions. Actually, RES has been shown to inhibit protein degradation in several in vitro models [9] by interfering with nuclear factor kappa beta (NF-kB) activation and nuclear translocation, inhibiting the signaling pathway that contributes to muscle mass loss [25]. Accordingly, RES supplementation has been shown to attenuate age-dependent fiber area decrease [26]. Interestingly, the administration of RES in rats undergoing hindlimb suspension did not prevent fiber atrophy during the period of disuse, but it increased the cross-sectional area of type II fibers in response to reloading, most probably by reducing pro-apoptotic signals [27].

2.2. Resveratrol and the NMJ

The NMJ is the point of communication between the motor neuron and the skeletal muscle cell, and it is the site for the transmission of action potential to activate contraction. The integrity of the NMJ is perturbed in neuromuscular disorders and in skeletal muscle aging and disuse [20,28–30], with a consequent loss of its organization, fragmentation and degeneration, contributing to reduced muscle performance. Already in 2006, Lagouge and collaborators [7], based on the evidence that RES improved the motor coordination and traction force in mice fed on a high-fat diet, suggested that RES could modulate neuromuscular communication. Later, analogously to the reported evidence that calorie restriction and exercise can attenuate age-dependent NMJ modifications [31], RES was reported to slow aging of the NMJ by reducing its fragmentation and denervation [32].

Although the molecular pathways regulating the NMJ-specific domain targeted by RES remain quite unexplored (the research in PubMed with the words "resveratrol neuromuscular junction" issued only five results), several research articles exploring RES effects on nervous tissue reveal significant neuroprotective action. Actually, in various experimental models, RES administration has been shown to reduce nerve cell senescence [33] and to ameliorate oxidative stress [34], endoplasmic reticulum stress [35] and inflammation [36], thereby improving locomotor function. This evidence suggests that RES has the potential to target the presynaptic component of the NMJ. Further work is needed to understand if and how RES can target the postsynaptic domain of the skeletal muscle cell.

2.3. Resveratrol, Connective Sheaths and Tendons

Connective sheaths and tendons are formed by an insoluble scaffold of the extracellular matrix (ECM), rich in collagen and elastic fibers, proteoglycans, glycoproteins and laminins, which is synthesized by fibroblasts. Connective sheaths serve as structural support for muscle fibers; participate in lateral and longitudinal force transmission; host immune system cells, satellite cells, nerves and capillaries; and form a strict connection with bones to displace the different parts of the skeleton [37].

In general, ECM homeostasis is maintained by the fine regulation of the production of its components and the activation of degrading enzymes [18]. In the context of a muscle, the presence of continuous excitation–contraction cycles requires an appropriate remodeling of the ECM, which can be highly impacted by exercise, age and pathological conditions [37]. For example, in aging individuals, the ECM undergoes fibrotic changes that lead to an increase in skeletal muscle stiffness, strength loss and injury predisposition [19,38]. Moreover, attesting to the crucial role of connective sheaths, muscle contraction can be impaired in ECM-specific disorders but also in apparently unrelated pathologies that result in connective tissue impairment [39]. For instance, patients affected by Ehlers–Danlos syndrome often display neuromuscular involvement [40]. Although these subjects experience skeletal muscle symptoms such as pain, fatigue and cramps, they have normal skeletal muscles, suggesting that muscle symptoms depend on the associated connective tissue [41]. Interestingly, patients affected by chronic kidney disease present reduced muscle performance accompanied by fibrosis, capillary rarefaction and weakness, which can be partially reverted by dialysis [39,42].

The effects of RES on the ECM are quite conflicting, which may be due to the different biochemical properties of the ECM in different tissues and the experimental models tested. It has been suggested that RES may modulate both the deposition and the degradation of the ECM components. RES has been shown to increase collagen deposition, improving wound healing and neovascularization after laparotomy in rats [43] but also to reduce cardiac fibrosis in several experimental models via the diminution of ECM component deposition and the regulation of metalloproteinase (MMP) activity [44]. In the context of skeletal muscle, there are few observations. Gliemann and collaborators reported that RES inhibits the training-induced expression of the metallopeptidase inhibitor-1 (TIMP-1) and reduces the levels of thrombospondin-1 (TSP-1), and they interpreted these data as an inhibition of the proangiogenic response and capillarization [45]. Considering the metalloproteinase-inhibiting function of TIMP-1 and the role of TSP-1 in promoting fibrosis [46], it could be hypothesized that RES plays an antifibrotic role also in the skeletal muscle.

Interestingly, because RES is a pleiotropic molecule and ECM is a highly elaborate and plastic tissue structure, modifications to connective sheaths could also depend on indirect RES actions, such as the scavenging of ROS, inhibition of the production of advanced glycation end products (AGEs), improvement of the inflammatory response and regulation of hormones [47].

2.4. Resveratrol and Skeletal Muscle Vascularization

In the skeletal muscle, microcirculation is responsible for the delivery of oxygen, nutrition and hormone molecules to and from muscle fibers and for the removal of heat and waste products [48]. To these aims, within the skeletal muscle, fiber type, fiber size, oxidative capacity and capillarization are strictly regulated to adapt to physiological needs [48–50]. Attesting to its crucial role, reduced capillarization impacts oxygen supply to muscles, contributing to exercise intolerance. Accordingly, poor capillarization levels are observed in hypertensive conditions, concurrent with metabolic dysregulation [51] and associated with lower muscle performance in older adults [52].

Based on the idea of the capillary bed as a target to improve skeletal muscle health, there is a growing interest in RES's potential to modulate capillarization, because studies in humans and animal models have provided promising results regarding both the skeletal muscle [53,54] and the cardiovascular system [55].

As suggested by Diaz and collaborators [56], the impact of RES on blood vessels is ascribable to a multilevel action, which starts from molecular regulation, passes through a biochemical response and converges to bring better blood supply to the tissue. RES has been shown to positively regulate vasculature through several mechanisms. It has been suggested that RES promotes angiogenesis via thioredoxin-1, heme oxygenase-1 and vascular endothelial growth factor (VEGF) [57] and regulates vasodilation by scavenging ROS and modulating nitric oxide synthesis. RES inhibits NFkB activation, leading to the reduction of inflammation markers and cytokines [53,58,59]. Ultimately, the improvement of capillarization in the skeletal muscle can also be credited with secondary effects, such as a decrease in fibrosis of the associated connective tissue [39,42] or an improvement in the cardiac hemodynamic properties, as shown in RES-treated diabetic rats [60].

3. Resveratrol and Skeletal Muscle in Clinical Trials

3.1. Search Strategy

We performed a systematic search from March 2023 to April 2023 following the Preferred Reporting Items for Systematic Reviews and Meta-Analyses (PRISMA) checklist guidelines. We searched the terms "resveratrol" and "skeletal muscle" in PubMed's clinical trials section, which returned 21 results. We searched the terms "resveratrol", "skeletal muscle" and "clinical trial" in Scopus, which returned 22 articles. We also searched the terms "resveratrol" and "skeletal muscle" in the Cochrane Library, and we obtained 53 results. We then selected only research articles that reported the analyses of skeletal muscle biopsies and/or physical performance. The cross-selection among libraries, together with a further analysis of the literature, returned 24 articles, which are reported in Table 1 [45,54,61–82]. Research articles from Gliemann et al. [45,78] and Olesen et al. [75], and research articles from Kjaer et al. [69] and Korsholm et al. [70], respectively, issued from a single clinical trial.

Table 1. List of the selected clinical trials reporting subjects' number, sex and age (reported between parentheses as mean age ± standard deviation), health condition, treatment conditions, analyses performed and results. Abbreviations: f, female; m, male; RES, resveratrol; PL, placebo; EGCG, epigallocatechine-3; T2DM, type 2 diabetes mellitus; COPD, chronic obstructive pulmonary disease; and PAD, peripheral arterial disease.

	Groups Subjects No./Sex/Age	Health Condition	Dose	Treatment	Treatment Length	Parameters Measured	Effect
SCOTTO DI PALUMBO, 2022 [61]	10 f + 11 m RES (75.0 ± 3.4) 8 f + 8 m PL (74.8 ± 3.9)	Sedentary	150 mg/day	Nutrition supplement drink	6 months	Performance	Positive effect
HARPER, 2021 [62]	13 f + 7 m RES 500 mg (72 ± 5.1) 18 f + 2 m RES 1000 mg (70.3 ± 5.8) 14 f + 6 m PL (73.3 ± 7.8)	Functional limitations	500 mg, 1000 mg/day	Resveratrol + exercise	12 weeks	Metabolism, inflammation markers, performance	Positive effect
HUANG, 2021 [63]	18 m RES (21.09 ± 1.33) 18 m PL (21.09 ± 1.33)	T2DM	500 mg, 1000 mg/day	Trans-resveratrol extract	7 days	Muscle damage markers	Positive effect
BEIJERS, 2020 [64]	4 f + 7 m RES (67.8 ± 9.0) 5 f + 5 m PL (65.3 ± 9.1)	COPD	150 mg/day	resVidaTM	4 weeks	Metabolism, inflammation markers	No effect
HUANG, 2020 [65]	18 m RES (20.2 ± 0.4) 18 m PL (20.2 ± 0.4)	Healthy	480 mg/day	Resveratrol extract + exercise	4 days	Metabolism markers	No effect
VAN POLANEN, 2020 [66]	18 m RES (61 ± 8) 18 m PL (61 ± 8)	Obese Normoglycemic	150 mg/day	Trans-resveratrol	30 days	Metabolism, histological markers	Positive effect
DE LIGT, 2018 [67]	13 m RES (66) 13 m PL (66)	Overweight (T2DM first-degree relatives)	150 mg/day	Trans-resveratrol	30–34 days	Metabolism markers, respiration	Partially positive effect

Table 1. *Cont.*

	Groups Subjects No./Sex/Age	Health Condition	Dose	Treatment	Treatment Length	Parameters Measured	Effect
ALWAY, 2017 [68]	9 f + 6 m RES (67.9 ± 1.1) 9 f + 6 m PL (67.9 ± 1.1)	Healthy	500 mg/day	Resveratrol + exercise	12 weeks	Metabolism, histological markers, performance	Positive effect
KJAER, 2017 [69]	21 m RES 150 mg (49.1 ± 1.46) 21 m RES 1000 mg (51.9 ± 1.28) 24 m PL (47.8 ± 1.3)	Metabolic syndrome	150 mg 1000 mg/day	Resveratrol	16 weeks	Metabolism markers	No effect
KORSHOLM, 2017 [70]	21 m RES (47.8 ± 1.3) 24 m PL (47.8 ± 1.3)	Metabolic syndrome	1000 mg/day	resVidaTM	4 months	Metabolomic analyses	Positive effect
MCDERMOTT, 2017 [71]	8 f + 13 m RES 125 mg (73.6 ± 6.6) 6 f + 17 m RES 500 mg (75.6 ± 7.3) 21 f + 45 m PL (74.4 ± 6.1)	PAD	125, 500 mg/day	Resveratrol	6 months	Performance	Partially positive effect
POLLACK, 2017 [54]	19 f (67 ± 7) 11 m (67 ± 7)	Glucose intolerant	2000–3000 mg/day	Resveratrol	6 weeks	Metabolism markers	Positive effect
MOST, 2016 [72]	18 EGCG + RES (36.1 ± 2.2) 20 PL (38.7 ± 2.2)	Overweight Obese	80 mg/day	Trans-resveratrol extract + epigallocatechine-3	12 weeks	Metabolism markers, respiration	Positive effects
POLLEY, 2016 [73]	4 f (19.7 ± 0.6) + 4 m (21.07 ± 2.4) RES 3 f (19.0 ± 0.8) + 5 m (20.0 ± 0.8) PL	Healthy	500 mg/day	Trans-resveratrol + 10mg/day piperine + exercise	4 weeks	Metabolism markers, respiration	Positive effects
GLIEMANN, 2014 [45]	9 m RES (60–72) 14 m RES + exercise (60–72) 7 m PL (60–72) 13 m PL + exercise (60–72)	Healthy	250 mg/day	Trans-resveratrol + exercise	8 weeks	Angiogenic markers	No effect
GOH, 2014 [74]	5 m RES (55.8 ± 7.3) 5 m PL (56.8 ± 5.3)	T2DM	500 mg/day	Trans-resveratrol	12 weeks	Metabolisms markers, performance	Partially positive effect
OLESEN, 2014 [75]	9 m RES (60–72) 14 m RES + exercise (60–72) 7 m PL (60–72) 13 m PL + exercise (60–72)	Healthy	250 mg/day	Trans-resveratrol + exercise	8 weeks	Metabolism, inflammation markers	No effect
SCRIBBANS, 2014 [76]	8 m RES (21 ± 1) 8 m PL (22 ± 1)	Healthy	150 mg/day	resVidaTM + exercise	4 weeks	Metabolism, histological markers, performance	Partially positive effect
WILLIAMS, 2014 [77]	8 m RES (23.8 ± 2.4) 8 m PL (23.4 ± 6.1)	Sedentary	300 mg	resVidaTM	Single dose	Metabolism markers, respiration	No effect

Table 1. *Cont.*

	Groups Subjects No./Sex/Age	Health Condition	Dose	Treatment	Treatment Length	Parameters Measured	Effect
GLIEMANN, 2013 [78]	9 m RES (60–72) 14 m RES + exercise (60–72) 7 m PL (60–72) 13 m PL + exercise (60–72)	Healthy	250 mg/day	Trans-resveratrol + exercise	8 weeks	Metabolism, inflammation markers	No effect
O'CONNOR, 2013 [79]	9 f + 11 m RES (19.9 ± 2.2) 12 f + 10 m PL (19.8 ± 1.7)	Healthy		Grape powder drink	45 days	Muscle performance, injury	No effect
POULSEN, 2013 [80]	12 m RES (44.7 ± 3.5) 12 m PL (31.1 ± 3.9)	Obese	1500 mg/day	Trans-resveratrol	4 weeks	Metabolism markers	No effect
YOSHINO, 2012 [81]	15 f RES (58.2.5 ± 4.0) 14 f PL (59.8.5 ± 4.3)	Healthy, postmenopausal	75 mg/day	Resveratrol	12 weeks	Metabolism markers	No effect
TIMMERS, 2011 [82]	11 m RES (52.5 ± 2.1) 11 m PL (52.5 ± 2.1)	Obese	150 mg/day	resVidaTM	30 days	Metabolism markers	Positive effect

3.2. *Effects of Resveratrol on Human Skeletal Muscle*

Table 1 reports the 24 research articles issuing from 21 clinical trials that were selected from the literature. Interestingly, 14 articles reported positive or partially positive effects upon RES treatment, while 10 articles reported no beneficial action.

At the first evaluation, the analysis of the selected papers brings to light the wide heterogeneity of the clinical trials in the design, settings, participants, interventions and main outcomes.

The number of subjects ranged from less than 10 to around 20. The mean age of the subjects, although being quite homogeneous in the distinct trials, varied from around 20 years to a maximum of 75 years (see Table 1). The reported clinical trials prevalently enrolled male subjects, with one protocol involving only women [81] and seven protocols involving both female and male subjects [61,62,64,68,71,73,79]. Enrolled subjects were healthy active and healthy sedentary [45,61,65,68,73,75–79,81]; aged persons with functional limitations [62]; and subjects affected by obesity and/or metabolic syndrome [66,67,69,70,72,80,82], peripheral arterial disease (PAD) [71], type 2 diabetes mellitus (T2DM) [74,83] and chronic obstructive pulmonary disease (COPD) [64]. In most of the protocols, RES was supplemented orally, alone in pills, in combination with epigallocatechine-3 (EGCG) [72] or piperine [73], as a drink [61,79] and associated with training protocols [62,65,68,73,75,76,78]. The daily dose ranged from a minimum of 75 to a maximum of 2000–3000 mg/day [54,81] in a single treatment or for periods of 4 days to 6 months [61,77] (see Table 1).

Although we selected only research articles that reported the analyses of skeletal muscle biopsies and/or physical performance, there are significant differences also in the parameters and methodologies exploited to evaluate RES effectiveness. These include protein and gene expression of muscle proteins and the specific variations in known RES targets such as SIRT-1, PGC1-alpha, mitochondrial metabolism, glucose-related metabolism, inflammation markers, systemic parameters, histological properties and physical function.

There is also a good degree of variability concerning the components of skeletal muscle investigated and relative outcomes. In brief, most of the articles focused on the characteristics of muscle fibers, investigating morphology, mitochondrial function, lipid or glycogen content, metabolic properties and inflammatory conditions. Some works reported a RES-dependent modulation of the signaling pathways involving AMPK, SIRT-1 and PGC-1-alpha [62,65,74], while others did not [64,75,76].

Some articles explored the effectiveness of RES on the modification of vascularization. Pollack and collaborators showed that RES improves vascular function [54], while Gliemann and collaborators reported that RES does not improve angiogenic response and capillary growth [45]. Gliemann and collaborators quantified also some markers of connective tissue integrity, finding that RES inhibits the training-induced expression of TIMP-1 and reduces the levels of TSP-1 [45]. The histological properties of the NMJ were not evaluated in the reported clinical trials. The analysis of physical performance was monitored in six clinical trials, which reported either a positive or partially positive effect or no effect.

4. Discussion

Collectively, the reported clinical trials do not provide a clear-cut picture of the effectiveness of RES in improving skeletal muscle health. RES action on skeletal muscle can be modulated by several factors, which can be summarized as the age, sex, physical activity and health conditions of the subjects; the dose and length of the treatment; and, not last, the influence of the complexity of the histological and physiological components of this organ.

There are some elementary considerations that can influence the outcome of a clinical protocol, such as the probability of dealing with subjects with different genetic background [84,85], the possibility of sex-dependent action [14] and the evidence that although in the same chronological age window, individuals age differently. This intrinsic variance in the sample can play an important role in the effectiveness of RES supplementation, as demonstrated in aging rats, where RES effects on vascular functions and biomarkers are age- and gender-dependent [86]. Moreover, the population heterogeneity may imply disparities in the baseline skeletal muscle characteristics, which, in turn, can produce widely distributed outcomes. Therefore, contrary to experimental models that provide quite homogeneous samples, well-suited for both cross-sectional and longitudinal investigations, we should keep in mind that the intrinsic heterogeneity of human samples could lead to important variability in the results, requiring prudence in the evaluation of the potentiality of RES.

Similarly, health and nutritional conditions of the enrolled subjects could be crucial to note when observing an effect of RES treatment. It has been proposed that discrepancies in PGC-1alpha and UPC3 regulation during RES supplementation in aging mice could be ascribable to feeding conditions: only mice fed on a high-fat diet [7,87] experienced a significant change in these two pathways. This evidence, together with the hypothesis that RES exerts measurable beneficial effects only when cell metabolism properties are compromised, could explain the lack of RES effect in human trials involving young healthy adults [76,79], and on the other side, why obese, overweight and T2DM young subjects benefited from RES supplementation even with short periods of treatment [65].

The length and the time window of the treatment seems to be crucial for a positive effect. Again, experimental models are very useful in interpreting data. In our previous work, we showed that while long-term treatment with RES was able to ameliorate skeletal muscle tissue inflammation and improve capillarization in aging mice [21,53], short-term treatment only improved the inflammatory conditions with no effect on the capillaries [59]. In this context, some clinical trials reported partially positive effects, showing improvement in a few crucial parameters or resulting in a high variability in the response to RES treatment [69,71,72]. We cannot exclude the possibility that longer treatments could be more beneficial.

The daily doses chosen in the distinct trials, which ranged from a minimum of 75 to a maximum of 2000–3000 mg/day, might be a source of diverging outcomes. Although it is possible that the observations are the results of a J-shaped dose–response effect, evidence from several studies suggests a complex dose-dependent action, where RES can selectively interact with diverse signaling pathways at different dose levels, leading to different physiological outcomes [88,89]. Moreover, variations in RES bioavailability, which depends on several factors such as the administration modalities and variability of the examined subjects, can be a crucial variable in the response to RES. Interestingly, recent research demonstrating that molecular modifications to RES can increase its bioactivity opens up the possibility of investigating different RES-derived compounds also in humans [90,91]. However, to date, the heterogeneity of the clinical trials reported makes it difficult to advance hypotheses on the optimal dosage and the use of strategies that can promote absorption.

Among its multiple effects, RES has been reported to act as an exercise-mimetic molecule. Accordingly, six clinical trials (see Table 1) investigated the effects of RES administered together with a training protocol. Three clinical trials reported positive effects [62,68,73], while three reported either no beneficial effects or unfavorable effects of RES treatment [45,65,75,76,78]. These clinical trials are quite heterogeneous with regard to the subjects involved, the dose, the treatment length, the training protocol applied, the parameters measured and the outcome (Table 1). However, in this case, the quality of exercise seems to be a discriminating factor for RES effectiveness. RES supplementation together with walking and whole-body resistance training [62], resistance training combined with aerobic training [68] and low-intensity submaximal training [73] resulted in positive effects, which include fiber hypertrophy [68], attenuation of the proinflammatory state [62] and mitochondrial capacity [73]. On the contrary, high-intensity interval training (HIIT) did not show beneficial outcomes [45,65,75,76,78]. Work from Gliemann, Olesen and collaborators showed that HIIT, but not RES, improves the metabolic parameters, inflammatory state and capillary growth in the skeletal muscle and ameliorates cardiovascular health [45,75,78]. Analogously, Scribbans and collaborators showed that 4 weeks of RES supplementation together with HIIT does not improve the maximal oxygen uptake and anaerobic exercise capacity and abolishes the exercise-dependent increase in the skeletal muscle expression of PGC-1, SIRT1 and super oxide dismutase 2 (SOD2). These data suggest that RES may somewhat impair the adaptive response to HIIT in humans [76].

In the evaluation of RES effectiveness, it is worth taking into consideration that the heterogeneity of the results could also depend on the analyses performed to evaluate RES action. As reported in Table 1, protocols envisaged the analysis of physical activity, metabolism, inflammation or other markers. The choice of which parameters to analyze could be a key factor in establishing RES effectiveness in a clinical trial. In our hands, the treatment of aging mice with RES improved the resistance to fatigue of isolated muscles but did not induce a significant improvement in performance in a treadmill test, which displayed a high variability among the tested animals [21]. Most probably, the lack of improvement in the treadmill test is ascribable to other physiological factors and components that participate in physical activity. This could apply to data obtained by Mc Dermott and collaborators [71], who reported that 125 mg/day of RES induces a statistically significant but not a clinically meaningful improvement in a 6 min walk test in subjects affected by PAD. It would be interesting to know whether other parameters (metabolism, signaling pathways, etc.) changed in subjects enrolled in the protocol.

Of note, we observed heterogeneity also in the methods of quantification of chosen parameters. The mitochondrial activity has been measured with near infrared spectroscopy (NIRS), ex vivo or via mitochondrial markers. Protein expression has been investigated by capillary nano-immunoassays [62], Western blot analysis [74] and/or mRNA levels [65,75,76]. We cannot exclude the possibility that the different methodologies could be a further source of discrepancy in studies that demonstrate differing results in protein expression levels (i.e., mitochondrial markers). In this context, the work performed by Kjaer, Korsholm and collaborators is of interest, which indicated that metabolomic analyses of tissues could evidence differences not perceivable by analyzing standard biochemical parameters [69,70]. In addition, the investigation of humans limits the possibility of detecting time-dependent or transitory changes, which could explain some of the divergences observed. In fact, contrary to cross-sectional protocols in animal models that allow one to study the progression of modifications, human subjects cannot undergo multiple muscle biopsies, and cross-sectional studies are prone to high variability and have an important impact on the outcome.

In the evaluation of the global effectiveness of RES on skeletal muscle contraction and physical performance, it should be taken into account that muscle contraction and physical performance depend on the coordination of distinct components of muscles, which possibly have different levels of susceptibility to RES. It can be suggested that the complexity of skeletal muscle physiology, together with individual, health and age-related variability, could profoundly impact the outcome of physical performance tests, despite the positive effect of RES on specific signaling pathways and/or histological and metabolic parameters. It is of note that in the clinical trials reported here, some crucial components of the skeletal muscles, such as the connective sheaths and the NMJ, were not investigated or were poorly investigated, demonstrating the need for further studies.

As a final thought, we cannot exclude the possibility that RES could induce differential effects in metabolic organs and systemic parameters [92] that do not result in a dramatic change in skeletal muscle characteristics but that can anyway contribute to the health of this tissue.

5. Conclusions

The present review reveals a broad heterogeneity in the protocols of RES administration and the analysis of its effects, which induces caution in the evaluation of its effectiveness. In order to reach a consensus, it would be beneficial to apply systematic programming of human clinical trials and consequent evaluative work. Moreover, given the broad spectrum of RES targets that are in strict correlation with the skeletal muscles, such as the cardiovascular system, endocrine system and digestive apparatus, it would be advantageous to perform multitargeted trials involving collaborative groups with distinct interests, know-how and expertise. The association of systematic planning and analysis together with collaboration among groups with distinct expertise could contribute to obtaining reliable results.

Author Contributions: Conceptualization, E.G. and L.T.; methodology, E.G. and L.T.; writing—original draft preparation, E.G.; writing—review and editing, E.G., M.C. and L.T. All authors have read and agreed to the published version of the manuscript.

Funding: This research received no external funding.

Institutional Review Board Statement: Not applicable.

Informed Consent Statement: Not applicable.

Data Availability Statement: Not applicable.

Conflicts of Interest: The authors declare no conflict of interest.

References

1. Cordova, A.C.; Jackson, L.S.M.; Berke-Schlessel, D.W.; Sumpio, B.E. The Cardiovascular Protective Effect of Red Wine. *J. Am. Coll. Surg.* **2005**, *200*, 428. [CrossRef]
2. Renaud, S.; de Lorgeril, M. Wine, Alcohol, Platelets, and the French Paradox for Coronary Heart Disease. *Lancet* **1992**, *339*, 1523–1526. [CrossRef]
3. Valenzano, D.R.; Terzibasi, E.; Genade, T.; Cattaneo, A.; Domenici, L.; Cellerino, A. Resveratrol Prolongs Lifespan and Retards the Onset of Age-Related Markers in a Short-Lived Vertebrate. *Curr. Biol.* **2006**, *16*, 296–300. [CrossRef]
4. Howitz, K.T.; Bitterman, K.J.; Cohen, H.Y.; Lamming, D.W.; Lavu, S.; Wood, J.G.; Zipkin, R.E.; Chung, P.; Kisielewski, A.; Zhang, L.-L.; et al. Small Molecule Activators of Sirtuins Extend Saccharomyces Cerevisiae Lifespan. *Nature* **2003**, *425*, 191–196. [CrossRef]
5. Wood, J.G.; Rogina, B.; Lavu, S.; Howitz, K.; Helfand, S.L.; Tatar, M.; Sinclair, D. Sirtuin Activators Mimic Caloric Restriction and Delay Ageing in Metazoans. *Nature* **2004**, *430*, 686–689. [CrossRef]
6. Chung, J.H.; Manganiello, V.; Dyck, J.R.B. Resveratrol as a Calorie Restriction Mimetic: Therapeutic Implications. *Trends Cell Biol.* **2012**, *22*, 546–554. [CrossRef]
7. Lagouge, M.; Argmann, C.; Gerhart-Hines, Z.; Meziane, H.; Lerin, C.; Daussin, F.; Messadeq, N.; Milne, J.; Lambert, P.; Elliott, P.; et al. Resveratrol Improves Mitochondrial Function and Protects against Metabolic Disease by Activating SIRT1 and PGC-1α. *Cell* **2006**, *127*, 1109–1122. [CrossRef]

8. Baur, J.A.; Pearson, K.J.; Price, N.L.; Jamieson, H.A.; Lerin, C.; Kalra, A.; Prabhu, V.V.; Allard, J.S.; Lopez-Lluch, G.; Lewis, K.; et al. Resveratrol Improves Health and Survival of Mice on a High-Calorie Diet. *Nature* **2006**, *444*, 337–342. [CrossRef]
9. Dirks Naylor, A.J. Cellular Effects of Resveratrol in Skeletal Muscle. *Life Sci.* **2009**, *84*, 637–640. [CrossRef]
10. Meng, X.; Zhou, J.; Zhao, C.-N.; Gan, R.-Y.; Li, H.-B. Health Benefits and Molecular Mechanisms of Resveratrol: A Narrative Review. *Foods* **2020**, *9*, 340. [CrossRef]
11. Jang, M.; Cai, L.; Udeani, G.O.; Slowing, K.V.; Thomas, C.F.; Beecher, C.W.W.; Fong, H.H.S.; Farnsworth, N.R.; Kinghorn, A.D.; Mehta, R.G.; et al. Cancer Chemopreventive Activity of Resveratrol, a Natural Product Derived from Grapes. *Science* **1997**, *275*, 218–220. [CrossRef]
12. Kulkarni, S.S.; Cantó, C. The Molecular Targets of Resveratrol. *Biochim. Biophys. Acta (BBA) Mol. Basis Dis.* **2015**, *1852*, 1114–1123. [CrossRef]
13. Pezzuto, J.M. The Phenomenon of Resveratrol: Redefining the Virtues of Promiscuity. *Ann. N. Y. Acad. Sci.* **2011**, *1215*, 123–130. [CrossRef]
14. Gehm, B.D.; McAndrews, J.M.; Chien, P.-Y.; Jameson, J.L. Resveratrol, a Polyphenolic Compound Found in Grapes and Wine, Is an Agonist for the Estrogen Receptor. *Proc. Natl. Acad. Sci. USA* **1997**, *94*, 14138–14143. [CrossRef]
15. Giacomello, E.; Toniolo, L. The Potential of Calorie Restriction and Calorie Restriction Mimetics in Delaying Aging: Focus on Experimental Models. *Nutrients* **2021**, *13*, 2346. [CrossRef]
16. Toniolo, L.; Giacomello, E. Resveratrol, Aging, and Fatigue. In *Aging: Oxidative Stress and Dietary Antioxidants*; Academic Press: Cambridge, MA, USA, 2020; pp. 309–317. ISBN 978-0-12-818698-5.
17. Tieland, M.; Trouwborst, I.; Clark, B.C. Skeletal Muscle Performance and Ageing. *J. Cachexia Sarcopenia Muscle* **2018**, *9*, 3–19. [CrossRef]
18. Kumar, L.; Bisen, M.; Khan, A.; Kumar, P.; Patel, S.K.S. Role of Matrix Metalloproteinases in Musculoskeletal Diseases. *Biomedicines* **2022**, *10*, 2477. [CrossRef]
19. Sinha, U.; Malis, V.; Chen, J.-S.; Csapo, R.; Kinugasa, R.; Narici, M.V.; Sinha, S. Role of the Extracellular Matrix in Loss of Muscle Force with Age and Unloading Using Magnetic Resonance Imaging, Biochemical Analysis, and Computational Models. *Front. Physiol.* **2020**, *11*, 626. [CrossRef]
20. Sirago, G.; Pellegrino, M.A.; Bottinelli, R.; Franchi, M.V.; Narici, M.V. Loss of Neuromuscular Junction Integrity and Muscle Atrophy in Skeletal Muscle Disuse. *Ageing Res. Rev.* **2023**, *83*, 101810. [CrossRef]
21. Toniolo, L.; Fusco, P.; Formoso, L.; Mazzi, A.; Canato, M.; Reggiani, C.; Giacomello, E. Resveratrol Treatment Reduces the Appearance of Tubular Aggregates and Improves the Resistance to Fatigue in Aging Mice Skeletal Muscles. *Exp. Gerontol.* **2018**, *111*, 170–179. [CrossRef]
22. Chi, T.-C.; Chen, W.-P.; Chi, T.-L.; Kuo, T.-F.; Lee, S.-S.; Cheng, J.-T.; Su, M.-J. Phosphatidylinositol-3-Kinase Is Involved in the Antihyperglycemic Effect Induced by Resveratrol in Streptozotocin-Induced Diabetic Rats. *Life Sci.* **2007**, *80*, 1713–1720. [CrossRef]
23. Su, H.-C.; Hung, L.-M.; Chen, J.-K. Resveratrol, a Red Wine Antioxidant, Possesses an Insulin-like Effect in Streptozotocin-Induced Diabetic Rats. *Am. J. Physiol.-Endocrinol. Metab.* **2006**, *290*, E1339–E1346. [CrossRef] [PubMed]
24. Szkudelski, T.; Szkudelska, K. Resveratrol and Diabetes: From Animal to Human Studies. *Biochim. Biophys. Acta (BBA) Mol. Basis Dis.* **2015**, *1852*, 1145–1154. [CrossRef]
25. Li, H.; Malhotra, S.; Kumar, A. Nuclear Factor-Kappa B Signaling in Skeletal Muscle Atrophy. *J. Mol. Med.* **2008**, *86*, 1113–1126. [CrossRef] [PubMed]
26. Hosoda, R.; Nakashima, R.; Yano, M.; Iwahara, N.; Asakura, S.; Nojima, I.; Saga, Y.; Kunimoto, R.; Horio, Y.; Kuno, A. Resveratrol, a SIRT1 Activator, Attenuates Aging-Associated Alterations in Skeletal Muscle and Heart in Mice. *J. Pharmacol. Sci.* **2023**, *152*, 112–122. [CrossRef] [PubMed]
27. Bennett, B.T.; Mohamed, J.S.; Alway, S.E. Effects of Resveratrol on the Recovery of Muscle Mass Following Disuse in the Plantaris Muscle of Aged Rats. *PLoS ONE* **2013**, *8*, e83518. [CrossRef] [PubMed]
28. Monti, E.; Reggiani, C.; Franchi, M.V.; Toniolo, L.; Sandri, M.; Armani, A.; Zampieri, S.; Giacomello, E.; Sarto, F.; Sirago, G.; et al. Neuromuscular Junction Instability and Altered Intracellular Calcium Handling as Early Determinants of Force Loss during Unloading in Humans. *J. Physiol.* **2021**, *599*, 3037–3061. [CrossRef] [PubMed]
29. Sirago, G.; Candia, J.; Franchi, M.V.; Sarto, F.; Monti, E.; Toniolo, L.; Reggiani, C.; Giacomello, E.; Zampieri, S.; Hartnell, L.M.; et al. Upregulation of Sarcolemmal Hemichannels and Inflammatory Transcripts with Neuromuscular Junction Instability during Lower Limb Unloading in Humans. *Biology* **2023**, *12*, 431. [CrossRef]
30. Gonzalez-Freire, M.; de Cabo, R.; Studenski, S.A.; Ferrucci, L. The Neuromuscular Junction: Aging at the Crossroad between Nerves and Muscle. *Front. Aging Neurosci.* **2014**, *6*, 208. [CrossRef]
31. Valdez, G.; Tapia, J.C.; Kang, H.; Clemenson, G.D.; Gage, F.H.; Lichtman, J.W.; Sanes, J.R. Attenuation of Age-Related Changes in Mouse Neuromuscular Synapses by Caloric Restriction and Exercise. *Proc. Natl. Acad. Sci. USA* **2010**, *107*, 14863–14868. [CrossRef]
32. Stockinger, J.; Maxwell, N.; Shapiro, D.; deCabo, R.; Valdez, G. Caloric Restriction Mimetics Slow Aging of Neuromuscular Synapses and Muscle Fibers. *J. Gerontol. A Biol. Sci. Med. Sci.* **2018**, *73*, 21–28. [CrossRef] [PubMed]

33. Liu, J.; Jiao, K.; Zhou, Q.; Yang, J.; Yang, K.; Hu, C.; Zhou, M.; Li, Z. Resveratrol Alleviates 27-Hydroxycholesterol-Induced Senescence in Nerve Cells and Affects Zebrafish Locomotor Behavior via Activation of SIRT1-Mediated STAT3 Signaling. *Oxid. Med. Cell Longev.* **2021**, *2021*, 6673343. [CrossRef] [PubMed]
34. Adedara, A.O.; Babalola, A.D.; Stephano, F.; Awogbindin, I.O.; Olopade, J.O.; Rocha, J.B.T.; Whitworth, A.J.; Abolaji, A.O. An Assessment of the Rescue Action of Resveratrol in Parkin Loss of Function-Induced Oxidative Stress in Drosophila Melanogaster. *Sci. Rep.* **2022**, *12*, 3922. [CrossRef]
35. Luo, Y.; Zhao, Y.; Lai, J.; Wei, L.; Zhou, G.; Yu, Y.; Liu, J. Resveratrol Suppresses Bupivacaine-Induced Spinal Neurotoxicity in Rats by Inhibiting Endoplasmic Reticulum Stress via SIRT1 Modulation. *Biomed. Res. Int.* **2023**, *2023*, 1176232. [CrossRef] [PubMed]
36. Zhao, H.; Mei, X.; Yang, D.; Tu, G. Resveratrol Inhibits Inflammation after Spinal Cord Injury via SIRT-1/NF-KB Signaling Pathway. *Neurosci. Lett.* **2021**, *762*, 136151. [CrossRef]
37. Csapo, R.; Gumpenberger, M.; Wessner, B. Skeletal Muscle Extracellular Matrix–What Do We Know About Its Composition, Regulation, and Physiological Roles? A Narrative Review. *Front. Physiol.* **2020**, *11*, 253. [CrossRef]
38. Lieber, R.L.; Ward, S.R. Cellular Mechanisms of Tissue Fibrosis. 4. Structural and Functional Consequences of Skeletal Muscle Fibrosis. *Am. J. Physiol.-Cell Physiol.* **2013**, *305*, C241–C252. [CrossRef]
39. Brightwell, C.R.; Kulkarni, A.S.; Paredes, W.; Zhang, K.; Perkins, J.B.; Gatlin, K.J.; Custodio, M.; Farooq, H.; Zaidi, B.; Pai, R.; et al. Muscle Fibrosis and Maladaptation Occur Progressively in CKD and Are Rescued by Dialysis. *JCI Insight* **2021**, *6*, e150112. [CrossRef]
40. Voermans, N.C.; van Alfen, N.; Pillen, S.; Lammens, M.; Schalkwijk, J.; Zwarts, M.J.; van Rooij, I.A.; Hamel, B.C.J.; van Engelen, B.G. Neuromuscular Involvement in Various Types of Ehlers–Danlos Syndrome. *Ann. Neurol.* **2009**, *65*, 687–697. [CrossRef]
41. Nygaard, R.H.; Jensen, J.K.; Voermans, N.C.; Heinemeier, K.M.; Schjerling, P.; Holm, L.; Agergaard, J.; Mackey, A.L.; Andersen, J.L.; Remvig, L.; et al. Skeletal Muscle Morphology, Protein Synthesis, and Gene Expression in Ehlers-Danlos Syndrome. *J. Appl. Physiol.* **2017**, *123*, 482–488. [CrossRef]
42. Abramowitz, M.K.; Paredes, W.; Zhang, K.; Brightwell, C.R.; Newsom, J.N.; Kwon, H.-J.; Custodio, M.; Buttar, R.S.; Farooq, H.; Zaidi, B.; et al. Skeletal Muscle Fibrosis Is Associated with Decreased Muscle Inflammation and Weakness in Patients with Chronic Kidney Disease. *Am. J. Physiol.-Ren. Physiol.* **2018**, *315*, F1658–F1669. [CrossRef] [PubMed]
43. Yaman, I.; Derici, H.; Kara, C.; Kamer, E.; Diniz, G.; Ortac, R.; Sayin, O. Effects of Resveratrol on Incisional Wound Healing in Rats. *Surg. Today* **2013**, *43*, 1433–1438. [CrossRef] [PubMed]
44. Yu, D.; Tang, Z.; Li, B.; Yu, J.; Li, W.; Liu, Z.; Tian, C. Resveratrol against Cardiac Fibrosis: Research Progress in Experimental Animal Models. *Molecules* **2021**, *26*, 6860. [CrossRef] [PubMed]
45. Gliemann, L.; Olesen, J.; Biensø, R.S.; Schmidt, J.F.; Akerstrom, T.; Nyberg, M.; Lindqvist, A.; Bangsbo, J.; Hellsten, Y. Resveratrol Modulates the Angiogenic Response to Exercise Training in Skeletal Muscles of Aged Men. *Am. J. Physiol.-Heart Circ. Physiol.* **2014**, *307*, H1111–H1119. [CrossRef] [PubMed]
46. Xu, L.; Zhang, Y.; Chen, J.; Xu, Y. Thrombospondin-1: A Key Protein That Induces Fibrosis in Diabetic Complications. *J. Diabetes Res.* **2020**, *2020*, e8043135. [CrossRef]
47. Agarwal, R.; Agarwal, P. Targeting Extracellular Matrix Remodeling in Disease: Could Resveratrol Be a Potential Candidate? *Exp. Biol. Med.* **2017**, *242*, 374–383. [CrossRef]
48. Hendrickse, P.; Degens, H. The Role of the Microcirculation in Muscle Function and Plasticity. *J. Muscle Res. Cell Motil.* **2019**, *40*, 127–140. [CrossRef]
49. Giacomello, E.; Crea, E.; Torelli, L.; Bergamo, A.; Reggiani, C.; Sava, G.; Toniolo, L. Age Dependent Modification of the Metabolic Profile of the Tibialis Anterior Muscle Fibers in C57BL/6J Mice. *Int. J. Mol. Sci.* **2020**, *21*, 3923. [CrossRef]
50. Barnouin, Y.; McPhee, J.S.; Butler-Browne, G.; Bosutti, A.; De Vito, G.; Jones, D.A.; Narici, M.; Behin, A.; Hogrel, J.; Degens, H. Coupling between Skeletal Muscle Fiber Size and Capillarization Is Maintained during Healthy Aging. *J. Cachexia Sarcopenia Muscle* **2017**, *8*, 647–659. [CrossRef]
51. Gueugneau, M.; Coudy-Gandilhon, C.; Meunier, B.; Combaret, L.; Taillandier, D.; Polge, C.; Attaix, D.; Roche, F.; Féasson, L.; Barthélémy, J.-C.; et al. Lower Skeletal Muscle Capillarization in Hypertensive Elderly Men. *Exp. Gerontol.* **2016**, *76*, 80–88. [CrossRef]
52. Prior, S.J.; Ryan, A.S.; Blumenthal, J.B.; Watson, J.M.; Katzel, L.I.; Goldberg, A.P. Sarcopenia Is Associated with Lower Skeletal Muscle Capillarization and Exercise Capacity in Older Adults. *J. Gerontol. A Biol. Sci. Med. Sci.* **2016**, *71*, 1096–1101. [CrossRef] [PubMed]
53. Toniolo, L.; Formoso, L.; Torelli, L.; Crea, E.; Bergamo, A.; Sava, G.; Giacomello, E. Long-Term Resveratrol Treatment Improves the Capillarization in the Skeletal Muscles of Ageing C57BL/6J Mice. *Int. J. Food Sci. Nutr.* **2021**, *72*, 37–44. [CrossRef] [PubMed]
54. Pollack, R.M.; Barzilai, N.; Anghel, V.; Kulkarni, A.S.; Golden, A.; O'Broin, P.; Sinclair, D.A.; Bonkowski, M.S.; Coleville, A.J.; Powell, D.; et al. Resveratrol Improves Vascular Function and Mitochondrial Number but Not Glucose Metabolism in Older Adults. *J. Gerontol. A Biol. Sci. Med. Sci.* **2017**, *72*, 1703–1709. [CrossRef] [PubMed]
55. Dolinsky, V.W.; Dyck, J.R.B. Experimental Studies of the Molecular Pathways Regulated by Exercise and Resveratrol in Heart, Skeletal Muscle and the Vasculature. *Molecules* **2014**, *19*, 14919–14947. [CrossRef]
56. Diaz, M.; Degens, H.; Vanhees, L.; Austin, C.; Azzawi, M. The Effects of Resveratrol on Aging Vessels. *Exp. Gerontol.* **2016**, *85*, 41–47. [CrossRef]

57. Kaga, S.; Zhan, L.; Matsumoto, M.; Maulik, N. Resveratrol Enhances Neovascularization in the Infarcted Rat Myocardium through the Induction of Thioredoxin-1, Heme Oxygenase-1 and Vascular Endothelial Growth Factor. *J. Mol. Cell. Cardiol.* **2005**, *39*, 813–822. [CrossRef]
58. Pearson, K.J.; Baur, J.A.; Lewis, K.N.; Peshkin, L.; Price, N.L.; Labinskyy, N.; Swindell, W.R.; Kamara, D.; Minor, R.K.; Perez, E.; et al. Resveratrol Delays Age-Related Deterioration and Mimics Transcriptional Aspects of Dietary Restriction without Extending Life Span. *Cell Metab.* **2008**, *8*, 157–168. [CrossRef]
59. Sirago, G.; Toniolo, L.; Crea, E.; Giacomello, E. A Short-Term Treatment with Resveratrol Improves the Inflammatory Conditions of Middle-Aged Mice Skeletal Muscles. *Int. J. Food Sci. Nutr.* **2022**, *73*, 630–637. [CrossRef]
60. Bresciani, L.; Calani, L.; Bocchi, L.; Delucchi, F.; Savi, M.; Ray, S.; Brighenti, F.; Stilli, D.; Del Rio, D. Bioaccumulation of Resveratrol Metabolites in Myocardial Tissue Is Dose-Time Dependent and Related to Cardiac Hemodynamics in Diabetic Rats. *Nutr. Metab. Cardiovasc. Dis.* **2014**, *24*, 408–415. [CrossRef]
61. Scotto di Palumbo, A.; McSwiney, F.T.; Hone, M.; McMorrow, A.M.; Lynch, G.; De Vito, G.; Egan, B. Effects of a Long Chain N-3 Polyunsaturated Fatty Acid-Rich Multi-Ingredient Nutrition Supplement on Body Composition and Physical Function in Older Adults with Low Skeletal Muscle Mass. *J. Diet. Suppl.* **2022**, *19*, 499–514. [CrossRef]
62. Harper, S.A.; Bassler, J.R.; Peramsetty, S.; Yang, Y.; Roberts, L.M.; Drummer, D.; Mankowski, R.T.; Leeuwenburgh, C.; Ricart, K.; Patel, R.P.; et al. Resveratrol and Exercise Combined to Treat Functional Limitations in Late Life: A Pilot Randomized Controlled Trial. *Exp. Gerontol.* **2021**, *143*, 111111. [CrossRef] [PubMed]
63. Huang, B.; Li, X.; Zhu, X. The Role of GM130 in Nervous System Diseases. *Front. Neurol.* **2021**, *12*, 743787. [CrossRef] [PubMed]
64. Beijers, R.J.H.C.G.; Gosker, H.R.; Sanders, K.J.C.; de Theije, C.; Kelders, M.; Clarke, G.; Cryan, J.F.; van den Borst, B.; Schols, A.M.W.J. Resveratrol and Metabolic Health in COPD: A Proof-of-Concept Randomized Controlled Trial. *Clin. Nutr.* **2020**, *39*, 2989–2997. [CrossRef] [PubMed]
65. Huang, C.-C.; Liu, C.-C.; Tsao, J.-P.; Hsu, C.-L.; Cheng, I.-S. Effects of Oral Resveratrol Supplementation on Glycogen Replenishment and Mitochondria Biogenesis in Exercised Human Skeletal Muscle. *Nutrients* **2020**, *12*, 3721. [CrossRef]
66. Van Polanen, N.; Zacharewicz, E.; de Ligt, M.; Timmers, S.; Moonen-Kornips, E.; Schaart, G.; Hoeks, J.; Schrauwen, P.; Hesselink, M.K.C. Resveratrol-induced Remodelling of Myocellular Lipid Stores: A Study in Metabolically Compromised Humans. *Physiol. Rep.* **2021**, *9*, e14692. [CrossRef]
67. De Ligt, M.; Bruls, Y.M.H.; Hansen, J.; Habets, M.-F.; Havekes, B.; Nascimento, E.B.M.; Moonen-Kornips, E.; Schaart, G.; Schrauwen-Hinderling, V.B.; van Marken Lichtenbelt, W.; et al. Resveratrol Improves Ex Vivo Mitochondrial Function but Does Not Affect Insulin Sensitivity or Brown Adipose Tissue in First Degree Relatives of Patients with Type 2 Diabetes. *Mol. Metab.* **2018**, *12*, 39–47. [CrossRef]
68. Alway, S.E.; McCrory, J.L.; Kearcher, K.; Vickers, A.; Frear, B.; Gilleland, D.L.; Bonner, D.E.; Thomas, J.M.; Donley, D.A.; Lively, M.W.; et al. Resveratrol Enhances Exercise-Induced Cellular and Functional Adaptations of Skeletal Muscle in Older Men and Women. *J. Gerontol. A Biol. Sci. Med. Sci.* **2017**, *72*, 1595–1606. [CrossRef]
69. Kjær, T.N.; Ornstrup, M.J.; Poulsen, M.M.; Stødkilde-Jørgensen, H.; Jessen, N.; Jørgensen, J.O.L.; Richelsen, B.; Pedersen, S.B. No Beneficial Effects of Resveratrol on the Metabolic Syndrome: A Randomized Placebo-Controlled Clinical Trial. *J. Clin. Endocrinol. Metab.* **2017**, *102*, 1642–1651. [CrossRef]
70. Korsholm, A.S.; Kjær, T.N.; Ornstrup, M.J.; Pedersen, S.B. Comprehensive Metabolomic Analysis in Blood, Urine, Fat, and Muscle in Men with Metabolic Syndrome: A Randomized, Placebo-Controlled Clinical Trial on the Effects of Resveratrol after Four Months' Treatment. *Int. J. Mol. Sci.* **2017**, *18*, 554. [CrossRef]
71. McDermott, M.M.; Leeuwenburgh, C.; Guralnik, J.M.; Tian, L.; Sufit, R.; Zhao, L.; Criqui, M.H.; Kibbe, M.R.; Stein, J.H.; Lloyd-Jones, D.; et al. Effect of Resveratrol on Walking Performance in Older People With Peripheral Artery Disease. *JAMA Cardiol.* **2017**, *2*, 902–907. [CrossRef]
72. Most, J.; Timmers, S.; Warnke, I.; Jocken, J.W.; van Boekschoten, M.; de Groot, P.; Bendik, I.; Schrauwen, P.; Goossens, G.H.; Blaak, E.E. Combined Epigallocatechin-3-Gallate and Resveratrol Supplementation for 12 Wk Increases Mitochondrial Capacity and Fat Oxidation, but Not Insulin Sensitivity, in Obese Humans: A Randomized Controlled Trial1,2. *Am. J. Clin. Nutr.* **2016**, *104*, 215–227. [CrossRef] [PubMed]
73. Polley, K.R.; Jenkins, N.; O'Connor, P.; McCully, K. Influence of Exercise Training with Resveratrol Supplementation on Skeletal Muscle Mitochondrial Capacity. *Appl. Physiol. Nutr. Metab.* **2016**, *41*, 26–32. [CrossRef] [PubMed]
74. Goh, K.P.; Lee, H.Y.; Lau, D.P.; Supaat, W.; Chan, Y.H.; Koh, A.F.Y. Effects of Resveratrol in Patients with Type 2 Diabetes Mellitus on Skeletal Muscle SIRT1 Expression and Energy Expenditure. *Int. J. Sport. Nutr. Exerc. Metab.* **2014**, *24*, 2–13. [CrossRef]
75. Olesen, J.; Gliemann, L.; Biensø, R.; Schmidt, J.; Hellsten, Y.; Pilegaard, H. Exercise Training, but Not Resveratrol, Improves Metabolic and Inflammatory Status in Skeletal Muscle of Aged Men. *J. Physiol.* **2014**, *592*, 1873–1886. [CrossRef]
76. Scribbans, T.D.; Ma, J.K.; Edgett, B.A.; Vorobej, K.A.; Mitchell, A.S.; Zelt, J.G.E.; Simpson, C.A.; Quadrilatero, J.; Gurd, B.J. Resveratrol Supplementation Does Not Augment Performance Adaptations or Fibre-Type–Specific Responses to High-Intensity Interval Training in Humans. *Appl. Physiol. Nutr. Metab.* **2014**, *39*, 1305–1313. [CrossRef] [PubMed]
77. Williams, C.B.; Hughes, M.C.; Edgett, B.A.; Scribbans, T.D.; Simpson, C.A.; Perry, C.G.R.; Gurd, B.J. An Examination of Resveratrol's Mechanisms of Action in Human Tissue: Impact of a Single Dose In Vivo and Dose Responses in Skeletal Muscle Ex Vivo. *PLoS ONE* **2014**, *9*, e102406. [CrossRef]

78. Gliemann, L.; Schmidt, J.F.; Olesen, J.; Biensø, R.S.; Peronard, S.L.; Grandjean, S.U.; Mortensen, S.P.; Nyberg, M.; Bangsbo, J.; Pilegaard, H.; et al. Resveratrol Blunts the Positive Effects of Exercise Training on Cardiovascular Health in Aged Men. *J. Physiol.* **2013**, *591*, 5047–5059. [CrossRef]
79. O'Connor, P.J.; Caravalho, A.L.; Freese, E.C.; Cureton, K.J. Grape Consumption's Effects on Fitness, Muscle Injury, Mood, and Perceived Health. *Int. J. Sport. Nutr. Exerc. Metab.* **2013**, *23*, 57–64. [CrossRef]
80. Poulsen, M.M.; Vestergaard, P.F.; Clasen, B.F.; Radko, Y.; Christensen, L.P.; Stødkilde-Jørgensen, H.; Møller, N.; Jessen, N.; Pedersen, S.B.; Jørgensen, J.O.L. High-Dose Resveratrol Supplementation in Obese Men: An Investigator-Initiated, Randomized, Placebo-Controlled Clinical Trial of Substrate Metabolism, Insulin Sensitivity, and Body Composition. *Diabetes* **2013**, *62*, 1186–1195. [CrossRef] [PubMed]
81. Yoshino, J.; Conte, C.; Fontana, L.; Mittendorfer, B.; Imai, S.; Schechtman, K.B.; Gu, C.; Kunz, I.; Fanelli, F.R.; Patterson, B.W.; et al. Resveratrol Supplementation Does Not Improve Metabolic Function in Non-Obese Women with Normal Glucose Tolerance. *Cell Metab.* **2012**, *16*, 658–664. [CrossRef]
82. Timmers, S.; Konings, E.; Bilet, L.; Houtkooper, R.H.; van de Weijer, T.; Goossens, G.H.; Hoeks, J.; van der Krieken, S.; Ryu, D.; Kersten, S.; et al. Calorie Restriction-like Effects of 30 Days of Resveratrol (ResVida^{TM}) Supplementation on Energy Metabolism and Metabolic Profile in Obese Humans. *Cell Metab.* **2011**, *14*, 612–622. [CrossRef] [PubMed]
83. Huang, C.-C.; Lee, M.-C.; Ho, C.-S.; Hsu, Y.-J.; Ho, C.-C.; Kan, N.-W. Protective and Recovery Effects of Resveratrol Supplementation on Exercise Performance and Muscle Damage Following Acute Plyometric Exercise. *Nutrients* **2021**, *13*, 3217. [CrossRef] [PubMed]
84. Murgia, M.; Brocca, L.; Monti, E.; Franchi, M.V.; Zwiebel, M.; Steigerwald, S.; Giacomello, E.; Sartori, R.; Zampieri, S.; Capovilla, G.; et al. Plasma Proteome Profiling of Healthy Subjects Undergoing Bed Rest Reveals Unloading-dependent Changes Linked to Muscle Atrophy. *J. Cachexia Sarcopenia Muscle* **2022**, *14*, 439–451. [CrossRef] [PubMed]
85. Sirago, G.; Picca, A.; Giacomello, E.; Marzetti, E.; Toniolo, L. The Contribution of Genetics to Muscle Disuse, Retraining, and Aging. *Genes* **2022**, *13*, 1378. [CrossRef] [PubMed]
86. Han, S.; Bal, N.B.; Sadi, G.; Usanmaz, S.E.; Uludag, M.O.; Demirel-Yilmaz, E. The Effects of Resveratrol and Exercise on Age and Gender-Dependent Alterations of Vascular Functions and Biomarkers. *Exp. Gerontol.* **2018**, *110*, 191–201. [CrossRef]
87. Barger, J.L.; Kayo, T.; Vann, J.M.; Arias, E.B.; Wang, J.; Hacker, T.A.; Wang, Y.; Raederstorff, D.; Morrow, J.D.; Leeuwenburgh, C.; et al. A Low Dose of Dietary Resveratrol Partially Mimics Caloric Restriction and Retards Aging Parameters in Mice. *PLoS ONE* **2008**, *3*, e2264. [CrossRef]
88. Bosutti, A.; Degens, H. The Impact of Resveratrol and Hydrogen Peroxide on Muscle Cell Plasticity Shows a Dose-Dependent Interaction. *Sci. Rep.* **2015**, *5*, 8093. [CrossRef]
89. Cho, S.-J.; Jung, U.J.; Choi, M.-S. Differential Effects of Low-Dose Resveratrol on Adiposity and Hepatic Steatosis in Diet-Induced Obese Mice. *Br. J. Nutr.* **2012**, *108*, 2166–2175. [CrossRef]
90. Nawaz, W.; Zhou, Z.; Deng, S.; Ma, X.; Ma, X.; Li, C.; Shu, X. Therapeutic Versatility of Resveratrol Derivatives. *Nutrients* **2017**, *9*, 1188. [CrossRef]
91. Biasutto, L.; Mattarei, A.; Azzolini, M.; La Spina, M.; Sassi, N.; Romio, M.; Paradisi, C.; Zoratti, M. Resveratrol Derivatives as a Pharmacological Tool. *Ann. N. Y. Acad. Sci.* **2017**, *1403*, 27–37. [CrossRef]
92. Svensson, K.; Schnyder, S.; Albert, V.; Cardel, B.; Quagliata, L.; Terracciano, L.M.; Handschin, C. Resveratrol and SRT1720 Elicit Differential Effects in Metabolic Organs and Modulate Systemic Parameters Independently of Skeletal Muscle Peroxisome Proliferator-Activated Receptor γ Co-Activator 1α (PGC-1α). *J. Biol. Chem.* **2015**, *290*, 16059–16076. [CrossRef] [PubMed]

nutrients

MDPI

Article

Foods, Nutrients, and Risk of In-Hospital Frailty in Women: Findings from a Large Prospective Cohort Study

Huifeng Zhang [1,2], Weimin Li [1,*], Youfa Wang [3,*], Yuanyuan Dong [2], Darren C. Greenwood [4], Laura J. Hardie [5] and Janet E. Cade [2]

1 Clinical Nutrition Department, The First Affiliated Hospital of Xi'an Jiaotong University, 277 Yanta West Road, Xi'an 710061, China; hfzhang@xjtu.edu.cn
2 Nutritional Epidemiology Group, School of Food Science and Nutrition, University of Leeds, Leeds LS2 9JT, UK; fsydo@leeds.ac.uk (Y.D.); j.e.cade@leeds.ac.uk (J.E.C.)
3 Global Health Institute, School of Public Health, Xi'an Jiaotong University, Xi'an 710116, China
4 Leeds Institute for Data Analytics, Faculty of Medicine and Health, University of Leeds, Leeds LS2 9JT, UK; d.c.greenwood@leeds.ac.uk
5 Division of Clinical and Population Sciences, Leeds Institute of Cardiovascular and Metabolic Medicine, School of Medicine, University of Leeds, Leeds LS2 9JT, UK; l.j.hardie@leeds.ac.uk
* Correspondence: liweimin1@xjtu.edu.cn (W.L.); youfawang@gmail.com (Y.W.)

Abstract: Frailty is increasingly prevalent worldwide because of aging populations. Diet may play a role as a modifiable risk factor. This study aimed to investigate associations between dietary factors and risk of frailty in the UK Women's Cohort admitted to hospitals in England. Consumption of foods and nutrients was estimated using a validated 217-item food frequency questionnaire at baseline. Incident frailty was assessed via a hospital frailty risk score based on linkage with hospital episode statistics. Out of 25,186 participants admitted to hospitals, 6919 (27%) were identified with frailty and 10,562 (42%) with pre-frailty over a mean follow-up of 12.7 years. After adjustment for confounding, we observed a 12% increase in risk of frailty with each additional 10 g/MJ intake of total meat (HR = 1.12, 95%CI: 1.07, 1.17), with the highest risk observed for processed meats (HR = 1.45, 95%CI: 1.21, 1.73). Similar associations were observed with pre-frailty. Vegetable intake was associated with slightly lower risk of frailty (HR = 0.98, 95%CI: 0.97, 1.00). There was no evidence of association between most nutrient intakes and in-hospital frailty risk. Overall, our findings suggest that reducing consumption of meat, especially processed meat, in adults may be beneficial regarding the development of frailty.

Keywords: frailty; dietary intakes; meat consumption; processed meat; nutrients; hospital episode statistics

1. Introduction

Frailty is a clinical syndrome characterized by cumulative loss of physiologic reserve and increased vulnerability to stressors. It is highly associated with increased risk of negative health-related outcomes, including falls, fractures, disability, and death [1]. The prevalence of frailty increases with age and rates are higher in women than in men [2,3]. Frailty is highly prevalent among hospital inpatients, with rates ranging from 4% to 89% in various settings [4]. A meta-analysis of over 450,000 geriatric hospital inpatients reported a prevalence of 47.4% for frailty and 73.2% for pre-frailty and frailty combined [5]. Frailty is an important risk factor for mortality and a broad range of adverse clinical outcomes, with the latter reflected in higher health-care use and expenditure [3,6,7]. It is reported that frail women have double the odds of hospitalization and three-fold increased odds of a nursing facility stay, and incur greater medical costs compared to robust women [8]. Given the health effects and rapidly growing aging population, frailty is increasingly gaining international attention as an important global health challenge [6].

A growing body of evidence points to a key role of nutrition in aging and age-related disease [9]. Dietary factors, encompassing dietary patterns, food groups, and intakes of calories, macro-nutrients, and micro-nutrients could be associated with the pathophysiology of frailty. Generally, healthier dietary patterns with high consumption of fruit, vegetables, and whole grains have been associated with lower frailty risk, both in cross-sectional and longitudinal studies [10]. However, significant heterogeneity is present across studies, and much of the evidence is either cross-sectional or limited in duration of follow-up. Although some longitudinal studies showed possible protective benefits from high consumption of fruit and vegetables [11,12] and low-fat dairy foods [13], and low consumption of ultra-processed foods [14], the number of studies is limited and the effects of these foods on the onset and progression of frailty are still unclear. There is a pressing need for better evidence to clarify potentially protective or harmful effects of food groups, nutrient intakes, and more on frailty risk.

As a modifiable lifestyle factor, nutrition could emerge as a potential target for future prevention and treatment strategies for frailty. This study aims to provide high-quality evidence for associations between diet and in-hospital frailty risk utilizing a large-scale population cohort, the UK Women's Cohort, with an extended period of follow-up.

2. Materials and Methods

2.1. Study Population

The UK Women's Cohort Study (UKWCS) has collected demographic, anthropometric, dietary, lifestyle, and health-related information from 35,372 women aged 35–69 years who responded to a postal questionnaire across England, Scotland, and Wales at baseline recruitment (1995 to 1998). The UKWCS was designed to explore potential associations between diet and chronic diseases, and has been described in detail elsewhere [15]. A subgroup of UKWCS participants who were admitted to hospitals in England were included in this study. A flow chart of UKWCS participants for this study is shown in Supplementary Figure S1.

At the cohort's inception in 1993, ethical approval was obtained from National Research Ethics Service (NRES) Committee for Yorkshire & the Humber—Leeds East (Ref: 15/YH/0027), and approval was updated to include linkage outcomes and related sub-studies (Health Research Authority, REC Reference: 17/YH/0144), along with an NHS Digital Data Sharing Agreement (DARS-NIC-109867-M8S6B-v1.5) for the UKWCS-HES database to include matched medical records.

2.2. Dietary Assessment

Dietary information at baseline was obtained from a self-administered 217-food item food frequency questionnaire (FFQ), which was adapted from the FFQ for the European Prospective Investigation into Cancer and Nutrition (EPIC) study [16]. This FFQ has been validated against four-day weighed food diaries and a second FFQ collected at the same time as the diary, involving 283 women 3 years after baseline. Whilst accepting that each tool type measures different aspects of diet, correlations between the two dietary assessment methods were comparable to those found in other studies [17,18]. The daily consumed food weight of each item (g/d) was calculated from daily portions multiplied by standard portion weights from the Food Standards Agency portion sizes book [19]. Daily portions were converted from food intake frequencies in the FFQ (details in Supplementary Table S1). Further, daily intakes of energy, macro-nutrients, and micro-nutrients were calculated by summing nutrient contents of each food item, from the consumed food weights multiplied by the standard nutrient composition of foods derived from McCance and Widdowson's The Composition of Foods (5th Edition) [20]. Nutrients provided by supplements were not included in this study.

In nutritional analyses of this study, the energy density method was used to model food and nutrient intakes [21]. Macro-nutrients (protein, carbohydrates, and fat) were included as the percentage of total energy derived from each one. Micro-nutrients and main foods were expressed as the intake per MJ of total energy. Then the multivariate energy density method was used in multiple regression models with daily energy intake as a covariate, as recommended by Willett et al. [21].

2.3. Outcome Variables

Incident cases of frailty were identified through a hospital frailty risk score (hFRS) developed and validated by Gilbert et al. [22]. Briefly, hFRS was based on International Classification of Diseases (ICD)-10 diagnostic codes that were the most related to frailty syndromes. The ICD-10 diagnostic codes in UKWCS came from linkage to hospital episode statistics (HES) of the UK National Health Service up to 31 March 2019. The HES contains multiple hospitalization records for each included participant with the main and secondary diagnostic ICD codes. The ICD codes listed in Supplementary Table S2 were assigned a corresponding point, while those not listed were given a point of 0. A hFRS for each participant was calculated by summing points from all diagnostic codes of one admitted hospital record. Participants were followed from study entry until first diagnosis of an event associated with frailty, date of death, or until the censor date (31 March 2019), whichever came first. Cases of pre-frailty or more severe frailty were defined if the hFRS > 0, and cases of frailty were defined if the hFRS ≥ 2.

2.4. Statistical Analysis

Descriptive statistics were used to summarize baseline socio-demographic, lifestyle, and nutritional characteristics for UKWCS participants within three groups of hFRS separately. Cox proportional hazards regression was used to estimate hazard ratios (HRs) and 95% confidence intervals (95% CIs) for associations between each dietary factor and risk of pre-frailty and frailty. For ease of interpretation, the HRs were presented per 10 g of the food groups per MJ of total energy consumed, as indicated in the results tables.

Unadjusted and multivariable-adjusted models were developed in this study. Potential risk factors for frailty previously identified in the literature were considered as covariates in the adjusted model, including age at baseline; ethnicity (white, Asian, black, and other); marital status (married/living as married, separated/divorced, and single/widowed); socio-economic status (SES, professional/managerial, intermediate, and routine/manual); physical activity (low, moderate, and high levels); smoking status (never smoked, ex-smoker, and current smoker); alcohol consumption (g/d, continuous); body mass index (BMI, kg/m^2, continuous); and daily energy intake (MJ/d, continuous). Most variables were self-reported at recruitment. SES was derived from the United Kingdom National Statistics-Socio-Economic Classification (NS-SEC) [23]. Physical activity was calculated based on a series of questions about participants' usual daily activities at baseline that were taken from the International Physical Activity Questionnaire (IPAQ) short form and categorized into three levels, being low, moderate, and high, according to the official guidelines for data processing and analysis [24].

To examine the effects relative to age on which frailty usually strongly depends, interaction terms between dietary factors and age were included in the Cox models, where age was used linearly. Additionally, subgroup analysis was conducted by fitting the models with participants <60 years old and $\geq$60 years old (at recruitment) separately. To further check for possible reverse causation, sensitivity analysis was conducted by excluding participants with survival time <3 years. A final sensitivity analysis was performed excluding participants aged <65 years at diagnosis to check for potential selection bias via inclusion of young age groups.

All statistical analyses were conducted using Stata/MP, version 17.0 (Stata Corp LP, College Station, TX, USA).

3. Results

3.1. Socioeconomic Characteristics and Dietary Intakes at Baseline

Of the 35,372 women at recruitment, 9980 women who did not have episode statistic hospital records and 206 women who had incomplete hospital records were excluded, leaving 25,186 women for analyses. Baseline characteristics of the participants admitted to hospitals are summarized according to hFRS levels in Table 1. After a mean follow-up of 12.7 years, there were 7705 women admitted to hospitals with no frailty (hFRS = 0), 10,562 women admitted to hospitals with pre-frailty (0 < hFRS < 2), and 6919 women admitted to hospitals with frailty (hFRS ≥ 2). On average, participants were 53.1 years old (standard deviation, SD = 9.4) at recruitment, and women with frailty or pre-frailty were older than women with no frailty. Consistent with this, women with frailty had an older mean age at diagnosis (64.1 years) compared with the other two groups (58.9 years hFRS = 0, 61.5 years 0 < hFRS < 2). Compared with participants with no frailty, those with frailty or pre-frailty had a higher proportion of no qualifications, were more likely to be single or widowed, and less likely to have professional or managerial jobs. The proportions of different ethnicity and levels of physical activity were similar across three groups. Body mass index was higher in women with pre-frailty (mean (SD) 24.8 (4.4) kg/m^2) and frailty (24.9 (4.5) kg/m^2) than women with no frailty (24.1 (3.9) kg/m^2). Women with no frailty reported drinking slightly more alcohol and smoking less than the other groups.

Table 1. Demographic characteristics of participants with different frailty scores at follow up in the UK Women's Cohort Study.

		hFRS = 0 (N = 7705, 30.6%)	**0 < hFRS < 2 (N = 10,562, 41.9%)**	**hFRS ≥ 2 (N = 6919, 27.5%)**	*p* *	**All Participants (N = 25,186)**
Age at baseline (years)	Mean (standard deviation)	49.4 (8.1)	54.0 (9.4)	56.0 (9.7)	<0.001	53.1 (9.4)
Age at diagnosis (years)	Mean (standard deviation)	58.9 (9.9)	61.5 (10.2)	64.1 (10.7)	<0.001	61.4 (10.5)
Follow-up time (years)	Mean (standard deviation)	14.4 (6.3)	11.7 (6.2)	12.3 (6.2)	<0.001	12.7 (6.3)
Ethnicity (N, %)	White	7447 (98.6%)	10,104 (98.5%)	6615 (98.8%)	0.346	24,166 (98.6%)
	Asian	45 (0.6%)	72 (0.7%)	38 (0.6%)		155 (0.6%)
	Black	13 (0.2%)	16 (0.2%)	11 (0.2%)		40 (0.2%)
	other	49 (0.7%)	66 (0.6%)	29 (0.4%)		144 (0.6%)
Educational level (N, %)	No qualifications	854 (11.8%)	1974 (20.8%)	1456 (23.8%)	<0.001	4284 (18.8%)
	O-level or equivalent	2404 (33.3%)	3097 (32.6%)	1796 (29.4%)		7297 (31.9%)
	A-level or equivalent	1825 (25.3%)	2184 (23.0%)	1409 (23.1%)		5418 (23.7%)
	University degree	2137 (29.6%)	2255 (23.7%)	1452 (23.8%)		5844 (25.6%)
Marital status (N, %)	Married or living as married	6023 (79.1%)	7675 (73.9%)	4890 (71.8%)	<0.001	18,588 (74.9%)
	Separated or divorced	804 (10.6%)	1158 (11.2%)	760 (11.2%)		2722 (11.0%)
	Single or widowed	787 (10.3%)	1556 (15.0%)	1162 (17.1%)		3505 (14.1%)
Socio-economic status (SES) (N, %)	Routine and manual	599 (7.9%)	1073 (10.4%)	707 (10.5%)	<0.001	2379 (9.7%)
	Intermediate	2086 (27.5%)	2912 (28.2%)	1984 (29.4%)		6982 (28.3%)
	Professional and managerial	4902 (64.6%)	6329 (61.4%)	4053 (60.1%)		15,284 (62.0%)
Physical activity (N, %)	Low level	791 (10.3%)	1211 (11.5%)	791 (11.4%)	0.178	2793 (11.1%)
	Moderate level	3918 (50.9%)	5205 (49.3%)	3379 (48.8%)		12,502 (49.6%)
	High level	2996 (38.9%)	4146 (39.3%)	2749 (39.7%)		9891 (39.3%)
Body mass index (BMI) (kg/m^2)	Mean (standard deviation)	24.1 (3.9)	24.8 (4.4)	24.9 (4.5)	<0.001	24.6 (4.3)
Alcohol (g/d)	Mean (standard deviation)	9.2 (10.3)	8.4 (10.5)	8.2 (10.0)	<0.001	8.6 (10.3)
Smoking status (N, %)	Never smoked	4477 (59.6%)	5707 (55.7%)	3745 (56.1%)	<0.001	13,929 (57.0%)
	Ex-smoker	2249 (30.0%)	3318 (32.4%)	2184 (32.7%)		7751 (31.7%)
	Current smoker	783 (10.4%)	1218 (11.9%)	742 (11.1%)		2743 (11.2%)

* Difference was tested using Student's *t*-test for continuous variables and $\chi 2$ test for categorical variables; hFRS, hospital frailty risk score.

Profiles of consumed main foods among participants admitted to hospitals in the UKWCS are summarized by frailty status in Table 2. Women with frailty consumed the highest absolute total fish, processed meat, red meat, and total meat, whilst women with no frailty consumed the lowest. Consumption of vegetables, fruits, and poultry was generally similar across the three groups. Dietary intakes of energy and nutrients in each group are summarized in Supplementary Table S3. Generally, there was little difference in daily intakes of energy and most nutrients at baseline across the three groups.

Table 2. Profiles of consumed main foods (g/day) among participants with different frailty scores at follow up in the UK Women's Cohort Study.

Food Groups	hFRS = 0 (N = 7705, 30.6%)	0 < hFRS < 2 (N = 10,562, 41.9%)	hFRS ≥ 2 (N = 6919, 27.5%)	p *	All Participants (N = 25,186)
Vegetables	316 (183)	322 (206)	318 (197)	0.120	319 (196)
Fruits	306 (226)	319 (256)	323 (251)	0.152	316 (246)
Total fish	27 (25)	29 (31)	30 (27)	<0.001	29 (28)
Processed meat	12 (14)	13 (16)	14 (16)	<0.001	13 (15)
Red meat	32 (38)	35 (44)	37 (47)	<0.001	34 (43)
Poultry	17 (20)	17 (21)	17 (22)	0.141	17 (21)
Total meat	63 (61)	67 (69)	70 (71)	<0.001	67 (67)

* Difference was tested by Student's t-test; hFRS, hospital frailty risk score.

3.2. Associations between Dietary Intakes and In-Hospital Frailty Risk

As shown in the upper panel of Figure 1, after adjustment for potential confounders, risk of pre- and more severe frailty (hFRS > 0) was 40% higher (HR = 1.40, 95%CI: 1.25, 1.56) with every additional 10 g/MJ of processed meat, 16% higher (HR = 1.16, 95%CI: 1.11, 1.21) per 10 g/MJ red meat, 8% higher (HR = 1.08, 95%CI: 1.00, 1.17) per 10 g/MJ poultry, and 10% higher (HR = 1.10, 95%CI: 1.07, 1.13) per 10 g/MJ total meat.

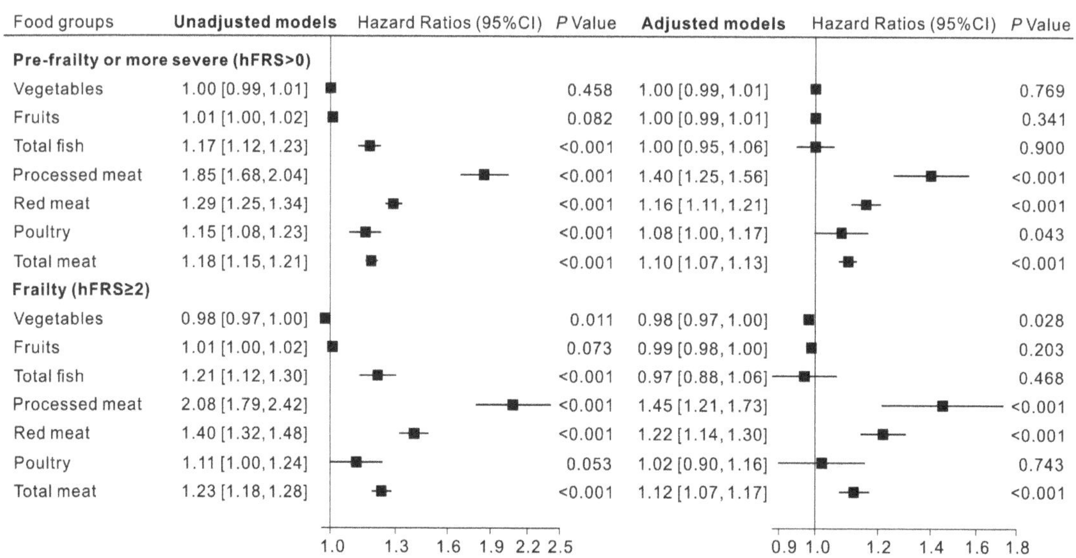

Figure 1. Associations between food groups (per 10 g/MJ) and in-hospital frailty risk within the UK Women's Cohort Study. Age, ethnicity, marital status, socioeconomic status, physical activity, body mass index, smoking status, alcohol consumption, and total energy intake were adjusted in the adjusted model in the right panel of the figure; 95%CI, 95% confidence interval; hFRS, hospital frailty risk score.

Similarly, for frailty (hFRS ≥ 2), as shown in the lower panel of Figure 1, risk was 45% higher (HR = 1.45, 95%CI: 1.21, 1.73) per 10 g/MJ processed meat, 22% higher (HR = 1.22,

95%CI: 1.14, 1.30) per 10 g/MJ red meat, 2% higher (HR = 1.02, 95%CI: 0.90, 1.16) per 10 g/MJ poultry, and 12% higher (HR = 1.12, 95%CI: 1.07, 1.17) per 10 g/MJ total meat.

Vegetable intake was associated with slightly lowered risk of frailty (hFRS $\geq$ 2) (HR = 0.98, 95%CI: 0.97, 1.00), but there was insufficient evidence of any association with risk of pre-frailty or more severe frailty (hFRS > 0) (HR = 1.00, 95%CI: 0.99, 1.01). There was no evidence of any association between consumption of fruits and total fish with risk of either pre-frailty or frailty.

Associations between intakes of energy and nutrients and in-hospital risk of pre-frailty or frailty are shown in the Supplementary Table S4. Daily intakes of vitamin B12 and zinc were associated with increased risk of pre-frailty or more severe frailty by 11% (HR = 1.11, 95%CI: 1.05, 1.17) and 12% (HR = 1.12, 95%CI: 1.04, 1.21), respectively, in adjusted models. All other nutrient intakes were not observed to be associated with frailty risk.

3.3. *Subgroup Analysis and Sensitivity Analysis*

For subgroup analysis in Table 3, the in-hospital risk of pre- and more severe frailty associated with consumption of processed meat, red meat, and total meat was higher in participants with age $\geq$60 years old compared to those with age <60 years old, where *p*-values for the interaction effect between age and each dietary factor were significant (0.015, 0.001, and <0.001 respectively). Similarly, the risk of frailty in relation to consumption of poultry and total meat was higher in participants with age $\geq$60 years old compared to those with age <60 years old, where *p*-interaction was 0.027 and 0.009, respectively. There was no significant evidence on effect modification of age on the remaining food groups (Table 3) and most nutrients (Supplementary Tables S5 and S6), except that a higher risk of frailty was observed in subjects with age $\geq$60 years old than those with age <60 years old associated with daily intake of zinc (*p*-interaction = 0.007 in Supplementary Table S5).

Table 3. Subgroup analysis by age on associations between food groups (per 10 g/MJ) and in-hospital frailty risk within the UK Women's Cohort Study.

	Hazard Ratio (95% Confidence Interval)				
	<60 Years Old		**≥60 Years Old**		**** *p*-Interaction With Age**
	Adjusted *	***p* ***	**Adjusted ***	***p* ***	
Risk of Pre- and more severe frailty (hFRS > 0)					
Vegetables	1.01 (0.99, 1.02)	0.383	0.99 (0.97, 1.01)	0.196	0.052
Fruits	1.00 (0.99, 1.01)	0.740	0.99 (0.97, 1.00)	0.083	0.154
Total fish	1.02 (0.95, 1.09)	0.614	0.99 (0.90, 1.09)	0.882	0.567
Processed meat	1.36 (1.18, 1.56)	<0.001	1.50 (1.23, 1.83)	<0.001	0.015
Red meat	1.14 (1.09, 1.21)	<0.001	1.19 (1.10, 1.28)	<0.001	0.001
Poultry	1.10 (1.00, 1.21)	0.042	1.07 (0.94, 1.22)	0.318	0.028
Total meat	1.09 (1.06, 1.13)	<0.001	1.13 (1.08, 1.19)	<0.001	<0.001
Risk of Frailty (hFRS ≥ 2)					
Vegetables	0.99 (0.98, 1.01)	0.591	0.97 (0.95, 0.99)	0.016	0.135
Fruits	1.00 (0.99, 1.02)	0.818	0.98 (0.96, 1.00)	0.038	0.708
Total fish	1.02 (0.91, 1.15)	0.681	0.90 (0.78, 1.04)	0.141	0.791
Processed meat	1.43 (1.14, 1.80)	0.002	1.54 (1.15, 2.06)	0.004	0.185
Red meat	1.24 (1.14, 1.35)	<0.001	1.20 (1.08, 1.34)	0.001	0.115
Poultry	1.01 (0.87, 1.19)	0.855	1.06 (0.87, 1.29)	0.586	0.027
Total meat	1.12 (1.06, 1.19)	<0.001	1.14 (1.05, 1.23)	0.001	0.009

hFRS, hospital frailty risk score; * adjusted for age, ethnicity, marital status, socioeconomic status, physical activity, body mass index, smoking status, alcohol consumption, and total energy intake; ** *p*-interaction represents the statistical significance for interaction item of dietary factor and age where age was modelled linearly in the Cox proportional regression.

In a sensitivity analysis, 3052 participants with a survival time <3 years were excluded to check for possible reverse causation. The risk of pre-frailty or frailty in relation to dietary factors appeared slightly attenuated, but did not change substantially after excluding those individuals (Supplementary Tables S7 and S8). In another sensitivity analysis, results were robust where participants (n = 15,601) aged <65 years at diagnosis were excluded (data shown in Supplementary Tables S9 and S10).

4. Discussion

This study found a significantly higher in-hospital risk of prefrailty or frailty associated with consumption of processed meat, red meat, and total meat. Subgroup analysis showed increased magnitude of these associations among individuals aged ≥60 years old compared to those <60 years old at baseline. Sensitivity analyses showed results were robust to the removal of participants with survival times <3 years in adjusted models.

Consumption of meat, especially red meat rich in pro-oxidative iron, is considered as part of a pro-inflammatory diet [25]. Analysis of a large, population-based cohort study of 455,776 participants in UK Biobank reported that a meat-based diet characterized by high consumption of red meat (lamb, pork, and beef) and processed meat was positively correlated with several pro-inflammatory biomarkers, including key leukocytes, C-reactive protein, and an aggregated inflammation score [26]. Recent studies have suggested a pro-inflammatory diet may be associated with increased risk of frailty [27,28]. A pro-inflammatory mechanism may provide a rationale for the positive association detected in this study between meat consumption and risk of frailty. Prospective evidence of frailty risk in relation to meat consumption is limited. Most previous studies tend to support a high intake of protein is recommended for the elderly to prevent frailty risk [29,30]. Although red meat and processed meat is rich in protein, we speculate that pro-inflammatory factors such as iron containing heme, saturated fat, and high levels of nitrates, nitrites, and amines may offset the protective effect of protein from meat. However, our findings need to be confirmed in other large longitudinal studies.

Most nutrient intakes were not observed to be associated with risk of frailty in our study and not consistent with mainstream opinion that protein supplementation, combined with physical activity, are an effective way to prevent physical frailty in elderly people [31,32]. We speculate a key reason for the inconsistency includes heterogeneity of frailty assessment tools. Previous studies have commonly used frailty tools based on phenotypes or deficits mainly including weight loss, exhaustion, slow gait speed, and weak grip strength [1], while our study used a hospital frailty score based on ICD diagnostic codes related to frailty syndromes, where frailty status is more likely to be severe. Heterogeneous assessment tools could assess different aspects of frailty, which may in turn modify the dietary associations detected. Intakes of vitamin B12 and zinc were found to be positively associated with risk of pre-frailty and frailty in our study. High levels of vitamin B12 are found to be associated with negative effects, such as inflammation and poor outcome for critically ill patients [33]. As reviewed, both deficient and high levels of vitamin B12 are risk factors for various clinical morbidities, and its levels potentially have an impact on frailty [34]. At present, there is little evidence on associations between intakes of zinc and risk of frailty. Generally, associations between nutrient intakes and frailty risk remain unconfirmed.

Currently, there is no uniform definition or assessment for the frailty complex. More than 60 assessment tools for frailty have been identified in scientific publications, of which nine are highly cited (≥200 citations), including the Fried Frailty Phenotype and the Rockwood Frailty Index [35]. The former, introduced by Fried et al., mainly comprises five phenotypes (weight loss, weakness, poor energy, slowness, low physical activity), where subjects having three or more phenotypes can be identified as frail [36]. The Rockwood Frailty Index is a broader deficit accumulation index, reflecting the proportion of potential deficits present in one person out of all considered deficits, including frailty symptoms, signs, diseases, and disabilities [37]. The hFRS tool has been validated against the two standard tests above and was used to assess the status of frailty in a hospital setting in this study [22]. Although most frailty assessment tools are more suitable for the elderly, our results were robust in sensitivity analysis to the exclusion of participants aged <65 years at diagnosis, indicating that the main results were not influenced substantially by inclusion of young participants in this study.

Strengths of this study include a large sample size and a longitudinal study design with relatively long follow-up time. In our study, the frailty assessment based on a cumulative score related to ICD codes from hospital records ensured identification accuracy and reduced reporting errors, as well as potential loss to follow-up over a long follow-up period. In addition, a variety of confounders including sociodemographic and lifestyle factors were adjusted for in our Cox proportional regression models. However, as for all observational studies, residual confounding is still possible. Moreover, as an observational study, causality cannot be established, although no obvious reverse causation was found in the sensitivity analysis of this study. In addition, taking hospital admission dates as a proxy for diagnosis dates of incident frailty could have resulted in measurement errors. Our study is also limited because only women admitted to hospitals in England were included in the analyses, which limits the generalizability of our findings; thus, more research is needed to investigate the risk of frailty in other populations.

In conclusion, our study revealed a link between in-hospital frailty risk and consumption of processed meat, red meat, and total meat. Further research is needed to elucidate the role of nutrition in strategies to reduce frailty. In particular, randomized controlled trials of plant-based protein as a meat substitute should be considered to provide high-quality evidence to support public health recommendations for preventing frailty.

Supplementary Materials: The following supporting information can be downloaded at: https://www.mdpi.com/article/10.3390/nu15214619/s1, Table S1: Transformation from intake frequencies to portion servings per day; Table S2: Frailty-related ICD-10 diagnostic codes derived from Gilbert et al. [22]; Table S3: Profiles of nutrient intakes by frailty status within the UK Women's Cohort Study; Table S4: Associations of nutrient intakes with in-hospital risk of pre-frailty and frailty within the UK Women's Cohort Study; Table S5: Subgroup analysis by age on associations between nutrient intakes and in-hospital risk of pre-frailty or more severe frailty within the UK Women's Cohort Study; Table S6: Subgroup analysis by age on associations between nutrient intakes and in-hospital risk of frailty within the UK Women's Cohort Study; Table S7: Sensitivity analysis to check for reverse causation on main foods within the UK Women's Cohort Study; Table S8: Sensitivity analysis to check for reverse causation on nutrient intakes within the UK Women's Cohort Study; Table S9: Sensitivity analysis excluding participants aged <65 y at diagnosis on main foods (per 10 g/MJ) within the UK Women's Cohort Study; Table S10: Sensitivity analysis excluding participants aged <65 y at diagnosis on nutrient intakes within the UK Women's Cohort Study.

Author Contributions: Conceptualization, H.Z., Y.W. and W.L.; methodology, H.Z., Y.W., W.L., L.J.H. and D.C.G.; software, H.Z. and Y.D.; validation, L.J.H., J.E.C. and D.C.G.; formal analysis, H.Z. and Y.D.; resources, J.E.C., D.C.G. and L.J.H.; data curation, J.E.C. and D.C.G.; writing—original draft preparation, H.Z.; writing—review and editing, L.J.H., D.C.G., J.E.C., W.L. and Y.W.; visualization, H.Z. and Y.D.; supervision, J.E.C., W.L. and Y.W.; project administration, H.Z. and J.E.C.; funding acquisition, H.Z. and J.E.C. All authors have read and agreed to the published version of the manuscript.

Funding: This research and the APC were funded by the Natural Science Basic Research Program of Shaanxi, grant number S2023-JC-QN-1424.

Institutional Review Board Statement: The study was conducted in accordance with the Declaration of Helsinki, and approved by National Research Ethics Service (NRES) Committee for Yorkshire & the Humber—Leeds East (Ref: 15/YH/0027), and was updated to include linkage outcomes and related sub-studies by Health Research Authority REC Reference: 17/YH/0144.

Informed Consent Statement: Informed consent has been obtained from all subjects involved in the study at recruitment of the cohort.

Data Availability Statement: Data is unavailable due to participants' privacy and ethical restrictions.

Acknowledgments: The authors thank the participants of the UK Women's Cohort Study for the provision of the data and the cohort team members who contributed to data collection. We also thank the World Cancer Research Fund (to J.E.C.) which funded the UK Women's Cohort.

Conflicts of Interest: The authors declare no conflict of interest. The funders had no role in the design of the study; in the collection, analyses, or interpretation of data; in the writing of the manuscript; or in the decision to publish the results.

References

1. Clegg, A.; Young, J.; Iliffe, S.; Rikkert, M.O.; Rockwood, K. Frailty in elderly people. *Lancet* **2013**, *381*, 752–762. [CrossRef] [PubMed]
2. Rohrmann, S. Epidemiology of Frailty in Older People. *Adv. Exp. Med. Biol.* **2020**, *1216*, 21–27. [CrossRef]
3. Hoogendijk, E.O.; Afilalo, J.; Ensrud, K.E.; Kowal, P.; Onder, G.; Fried, L.P. Frailty: Implications for clinical practice and public health. *Lancet* **2019**, *394*, 1365–1375. [CrossRef] [PubMed]
4. Veronese, N.; Custodero, C.; Cella, A.; Demurtas, J.; Zora, S.; Maggi, S.; Barbagallo, M.; Sabbà, C.; Ferrucci, L.; Pilotto, A. Prevalence of multidimensional frailty and pre-frailty in older people in different settings: A systematic review and meta-analysis. *Ageing Res. Rev.* **2021**, *72*, 101498. [CrossRef] [PubMed]
5. Doody, P.; Asamane, E.A.; Aunger, J.A.; Swales, B.; Lord, J.M.; Greig, C.A.; Whittaker, A.C. The prevalence of frailty and pre-frailty among geriatric hospital inpatients and its association with economic prosperity and healthcare expenditure: A systematic review and meta-analysis of 467,779 geriatric hospital inpatients. *Ageing Res. Rev.* **2022**, *80*, 101666. [CrossRef] [PubMed]
6. Dent, E.; Martin, F.C.; Bergman, H.; Woo, J.; Romero-Ortuno, R.; Walston, J.D. Management of frailty: Opportunities, challenges, and future directions. *Lancet* **2019**, *394*, 1376–1386. [CrossRef]
7. The Lancet. Bringing frailty into all realms of medicine. *Lancet* **2019**, *394*, 1298. [CrossRef]
8. Ensrud, K.E.; Kats, A.M.; Schousboe, J.T.; Taylor, B.C.; Cawthon, P.M.; Hillier, T.A.; Yaffe, K.; Cummings, S.R.; Cauley, J.A.; Langsetmo, L. Frailty Phenotype and Healthcare Costs and Utilization in Older Women. *J. Am. Geriatr. Soc.* **2018**, *66*, 1276–1283. [CrossRef]
9. Longo, V.D.; Anderson, R.M. Nutrition, longevity and disease: From molecular mechanisms to interventions. *Cell* **2022**, *185*, 1455–1470. [CrossRef]
10. Rashidi Pour Fard, N.; Amirabdollahian, F.; Haghighatdoost, F. Dietary patterns and frailty: A systematic review and meta-analysis. *Nutr. Rev.* **2019**, *77*, 498–513. [CrossRef]
11. Fung, T.T.; Struijk, E.A.; Rodriguez-Artalejo, F.; Willett, W.C.; Lopez-Garcia, E. Fruit and vegetable intake and risk of frailty in women 60 years old or older. *Am. J. Clin. Nutr.* **2020**, *112*, 1540–1546. [CrossRef] [PubMed]
12. Kojima, G.; Iliffe, S.; Jivraj, S.; Walters, K. Fruit and Vegetable Consumption and Incident Prefrailty and Frailty in Community-Dwelling Older People: The English Longitudinal Study of Ageing. *Nutrients* **2020**, *12*, 3882. [CrossRef] [PubMed]
13. Cuesta-Triana, F.; Verdejo-Bravo, C.; Fernández-Pérez, C.; Martín-Sánchez, F.J. Effect of Milk and Other Dairy Products on the Risk of Frailty, Sarcopenia, and Cognitive Performance Decline in the Elderly: A Systematic Review. *Adv. Nutr.* **2019**, *10*, S105–S119. [CrossRef] [PubMed]
14. Sandoval-Insausti, H.; Blanco-Rojo, R.; Graciani, A.; López-García, E.; Moreno-Franco, B.; Laclaustra, M.; Donat-Vargas, C.; Ordovás, J.M.; Rodríguez-Artalejo, F.; Guallar-Castillón, P. Ultra-processed Food Consumption and Incident Frailty: A Prospective Cohort Study of Older Adults. *J. Gerontol. A Biol. Sci. Med. Sci.* **2020**, *75*, 1126–1133. [CrossRef]
15. Cade, J.E.; Burley, V.J.; Alwan, N.A.; Hutchinson, J.; Hancock, N.; Morris, M.A.; Threapleton, D.E.; Greenwood, D.C. Cohort Profile: The UK Women's Cohort Study (UKWCS). *Int. J. Epidemiol.* **2017**, *46*, e11. [CrossRef]
16. Riboli, E.; Kaaks, R. The EPIC Project: Rationale and study design. European Prospective Investigation into Cancer and Nutrition. *Int. J. Epidemiol.* **1997**, *26* (Suppl. S1), S6–S14. [CrossRef]
17. Cade, J.E.; Burley, V.J.; Greenwood, D.C. The UK Women's Cohort Study: Comparison of vegetarians, fish-eaters and meat-eaters. *Public. Health Nutr.* **2004**, *7*, 871–878. [CrossRef]
18. Spence, M.; Cade, J.; Burley, V.; Greenwood, D. Ability of the UK Women's Cohort Study Food Frequency Questionnaire to rank dietary intakes: A preliminary validation study. *Proc. Nutr. Soc. Lond.* **2002**, *61*, 117A.
19. Food Standards Agency (FSA). *Food Portion Sizes*, 3rd ed.; Mills, A., Patel, S., Crawley, H., Eds.; The Stationary Office: London, UK, 2002.
20. Holland, B.; Welch, A.A.; Unwin, I.D.; Buss, D.H.; Paul, A.A.; Southgate, D. *McCance and Widdowson's the Composition of Foods*, 5th ed.; London Royal Society of Chemistry and MAFF: London, UK, 1991.
21. Willett, W. *Nutritional Epidemiology*, 3rd ed.; Oxford University Press: Oxford, UK, 2012; Volume 40.
22. Gilbert, T.; Neuburger, J.; Kraindler, J.; Keeble, E.; Smith, P.; Ariti, C.; Arora, S.; Street, A.; Parker, S.; Roberts, H.C.; et al. Development and validation of a Hospital Frailty Risk Score focusing on older people in acute care settings using electronic hospital records: An observational study. *Lancet* **2018**, *391*, 1775–1782. [CrossRef]
23. The National Statistics Socio-Economic Classification User Manual Office for National Statistics. Available online: https://www.ons.gov.uk/ons/guide-method/classifications/archived-standard-classifications/soc-and-sec-archive/the-national-statistics-socio-economic-classification{-}{-}user-manual.pdf (accessed on 28 July 2023).

24. Guidelines for Data Processing and Analysis of the International Physical Activity Questionnaire (IPAQ)-Short and Long Forms. IPAQ Research Committee 2005. Available online: http://www.ipaq.ki.se/scoring.pdf (accessed on 28 July 2023).
25. Rodríguez-Molinero, J.; Miguelánez-Medrán, B.D.C.; Puente-Gutiérrez, C.; Delgado-Somolinos, E.; Martín Carreras-Presas, C.; Fernández-Farhall, J.; López-Sánchez, A.F. Association between Oral Cancer and Diet: An Update. *Nutrients* **2021**, *13*, 1299. [CrossRef]
26. Shi, H.; Schweren, L.J.S.; Ter Horst, R.; Bloemendaal, M.; van Rooij, D.; Vasquez, A.A.; Hartman, C.A.; Buitelaar, J.K. Low-grade inflammation as mediator between diet and behavioral disinhibition: A UK Biobank study. *Brain Behav. Immun.* **2022**, *106*, 100–110. [CrossRef] [PubMed]
27. Moradi, S.; Hadi, A.; Mohammadi, H.; Asbaghi, O.; Zobeiri, M.; Marx, W.; Entezari, M.H. Dietary Inflammatory Index and the Risk of Frailty Among Older Adults: A Systematic Review and Meta-Analysis. *Res. Aging* **2021**, *43*, 323–331. [CrossRef]
28. Vicente, B.M.; Lucio Dos Santos Quaresma, M.V.; Maria de Melo, C.; Lima Ribeiro, S.M. The dietary inflammatory index (DII®) and its association with cognition, frailty, and risk of disabilities in older adults: A systematic review. *Clin. Nutr. ESPEN* **2020**, *40*, 7–16. [CrossRef] [PubMed]
29. Feart, C. Nutrition and frailty: Current knowledge. *Prog. Neuropsychopharmacol. Biol. Psychiatry* **2019**, *95*, 109703. [CrossRef] [PubMed]
30. Gimeno-Mallench, L.; Sanchez-Morate, E.; Parejo-Pedrajas, S.; Mas-Bargues, C.; Inglés, M.; Sanz-Ros, J.; Román-Domínguez, A.; Olaso, G.; Stromsnes, K.; Gambini, J. The Relationship between Diet and Frailty in Aging. *Endocr. Metab. Immune Disord. Drug Targets* **2020**, *20*, 1373–1382. [CrossRef] [PubMed]
31. Fiatarone, M.A.; O'Neill, E.F.; Ryan, N.D.; Clements, K.M.; Solares, G.R.; Nelson, M.E.; Roberts, S.B.; Kehayias, J.J.; Lipsitz, L.A.; Evans, W.J. Exercise training and nutritional supplementation for physical frailty in very elderly people. *N. Engl. J. Med.* **1994**, *330*, 1769–1775. [CrossRef]
32. O'Connell, M.L.; Coppinger, T.; McCarthy, A.L. The role of nutrition and physical activity in frailty: A review. *Clin. Nutr. ESPEN* **2020**, *35*, 1–11. [CrossRef]
33. Romain, M.; Sviri, S.; Linton, D.M.; Stav, I.; van Heerden, P.V. The role of Vitamin B12 in the critically ill—A review. *Anaesth. Intensive Care* **2016**, *44*, 447–452. [CrossRef]
34. Pyrgioti, E.E.; Karakousis, N.D. B12 levels and frailty syndrome. *J. Frailty Sarcopenia Falls* **2022**, *7*, 32–37. [CrossRef]
35. Buta, B.J.; Walston, J.D.; Godino, J.G.; Park, M.; Kalyani, R.R.; Xue, Q.L.; Bandeen-Roche, K.; Varadhan, R. Frailty assessment instruments: Systematic characterization of the uses and contexts of highly-cited instruments. *Ageing Res. Rev.* **2016**, *26*, 53–61. [CrossRef]
36. Fried, L.P.; Tangen, C.M.; Walston, J.; Newman, A.B.; Hirsch, C.; Gottdiener, J.; Seeman, T.; Tracy, R.; Kop, W.J.; Burke, G.; et al. Frailty in older adults: Evidence for a phenotype. *J. Gerontol. A Biol. Sci. Med. Sci.* **2001**, *56*, M146–M156. [CrossRef] [PubMed]
37. Rockwood, K.; Mitnitski, A. Frailty in relation to the accumulation of deficits. *J. Gerontol. A Biol. Sci. Med. Sci.* **2007**, *62*, 722–727. [CrossRef] [PubMed]

MDPI AG
Grosspeteranlage 5
4052 Basel
Switzerland
Tel.: +41 61 683 77 34

Nutrients Editorial Office
E-mail: nutrients@mdpi.com
www.mdpi.com/journal/nutrients